Consul: Up and Running
Service Mesh for Any Runtime or Cloud

Luke Kysow

Foreword by
Mitchell Hashimoto
and Armon Dadgar

Beijing · Boston · Farnham · Sebastopol · Tokyo

Consul: Up and Running

by Luke Kysow

Published by O'Reilly Media, Inc., 1005 Gravenstein Highway North, Sebastopol, CA 95472.

O'Reilly books may be purchased for educational, business, or sales promotional use. Online editions are also available for most titles (*http://oreilly.com*). For more information, contact our corporate/institutional sales department: 800-998-9938 or *corporate@oreilly.com*.

Acquisitions Editor: John Devins	**Indexer:** nSight, Inc.
Development Editor: Corbin Collins	**Interior Designer:** David Futato
Production Editor: Gregory Hyman	**Cover Designer:** Karen Montgomery
Copyeditor: Liz Wheeler	**Illustrator:** Kate Dullea
Proofreader: nSight, Inc.	

June 2022:   First Edition

Revision History for the First Edition

2022-06-01:   First Release

See *http://oreilly.com/catalog/errata.csp?isbn=9781098106140* for release details.

978-1-098-10614-0

[LSI]

To Isha, Kate, Mom, and Dad
For your love and support

Table of Contents

Foreword

We started HashiCorp in 2012 to solve the challenges introduced by the rise of public cloud. The products we introduced were built in roughly the same order in which they'd be experienced by a new team building in the cloud. Vagrant was first, since creating a development environment was the first challenge we faced. Packer was second, to translate those development environments into cloud images. With the applications successfully deployed, the next challenge was networking between the multiple deployed images, and thus Consul was born.

The word *microservice* wasn't used then, and the problem space was admittedly much smaller: we needed a way to find the address of a healthy instance of another application or service. The reality of cloud introduced several new technical challenges: global availability, automation friendliness, and the expectation that application instances came up and went down constantly. The initial release of Consul in 2014 solved all these challenges.

The beauty of new paradigms is that first-order challenges—once solved—give rise to second-order capabilities. The first-order challenge was service discovery in the world of public cloud. The second-order capability was then microservices, improved monitoring, more dynamic routing, and enhanced security by leveraging this new software-driven networking layer.

Consul was the natural place to enable these new capabilities, and over the years Consul has evolved to solve these difficult, modern networking challenges. From latency-aware routing at the DNS layer to automatic TLS between services to HTTP-aware load balancing and more, Consul has grown into a fully featured service mesh.

These capabilities enable teams to take full advantage of what public cloud has to offer while simultaneously getting more out of on-premises environments. Teams can deliver more applications across more regions safely, and teams that use multiple cloud platforms or on-premises datacenters can communicate across those environments in a consistent manner. And this is all possible without any modifications to the deployed applications.

Today, Consul is downloaded millions of times per year and is deployed into everything from small hobbyist home labs to the infrastructure of the world's largest companies. It has been proven in challenging production environments time and time again.

Luke Kysow has been part of the Consul engineering team for many years, personally implementing many of its incredible features. He has a particular talent in making complex topics approachable by anyone, and he does so beautifully in this book.

— Mitchell Hashimoto and Armon Dadgar
Cofounders of HashiCorp and creators of Consul

Preface

The sheer volume of software required by today's world has necessitated an evolution in how we structure our engineering organizations. We've learned that smaller, independent teams work better than larger, highly coupled ones. Since Conway's law—that companies will produce systems to match their organizational structure—is inevitable, this evolution has precipitated the rise of microservices: smaller, independent services owned by smaller, independent teams. As a result of these forces, companies are now running hundreds and even thousands of services in production.

The rise of microservices has enabled development teams to scale up and ship code faster, but it has also caused an exponential increase in complexity for operations teams. What was once an in-memory function call is now a cross-continent API request that can fail in unexpected and spectacular ways. What was once a single monitoring dashboard is now a byzantine maze of metrics, logs, and traces. A security model that was once a simple firewall now must protect against a myriad of ever-evolving attack vectors and threats. Finally, what was once a single monolithic service is now hundreds of services built using different technologies and deployed on multiple runtimes: virtual machines (VMs), Kubernetes, serverless platforms, and more.

Operations teams, also known as DevOps and site reliability engineering (SRE), thus face a monumental challenge. In the midst of this complexity, they must harden security, increase reliability, simplify observability, and speed application delivery—and they must do so in a way that works across multiple runtimes and languages. Service mesh is an exciting new technology that promises a solution to these problems.

Consul is a fully featured service mesh from HashiCorp, the company that also created Terraform, Vault, Nomad, Packer, and Vagrant. A small operations team can leverage Consul to impact security, reliability, observability, and application delivery across their entire stack—all without requiring developers to modify their underlying microservices.

In this book, you'll learn to install, configure, and operate Consul in order to tame complexity and take back control of your infrastructure. I'm excited for you to start on your service mesh journey with Consul—let's dig in and get up and running!

Who Should Read This Book

If you're a platform or operations engineer tasked with maintaining a growing micro-services environment on Kubernetes or VMs, then this book is for you.

If you're a microservices developer interested in increasing reliability or experimenting with advanced deployment strategies such as blue/green and canarying, this book is also for you.

Or perhaps your organization is already using Consul and you're looking to learn how it works at a deeper level and how to utilize it better.

This book will also be helpful for security engineers and higher-level decision-makers (managers, directors, VPs of engineering, and CTOs) to provide an overview of the concepts behind a service mesh and the value it provides.

This book assumes general knowledge of microservices development and networking concepts such as load balancers. It contains instructions for installing Consul on Kubernetes or Linux VMs and assumes that you will be familiar with one of those platforms. It contains exercises that you can complete on Windows, macOS, or Linux machines.

Navigating This Book

The book starts with service mesh fundamentals: what a service mesh is and how it works. Next, you'll learn what makes Consul unique, its architecture, and the specific protocols it uses. With that groundwork in place, you'll be ready to deploy Consul onto Kubernetes or VMs and add your services into the service mesh.

You'll then learn to use Consul to secure your systems with zero trust networking, add observability, increase reliability, and control traffic. In the final chapter, I cover advanced topics such as multi-cluster deployment.

Throughout the book, I include exercises for both Kubernetes and VMs, so you can utilize these concepts with an actual microservices application. If you wish to follow along with the exercises, I recommend you complete the chapters in order since they often rely on one another.

Join the official *Consul: Up and Running* Discord server (*https://discord.gg/zxQcUVYKeS*) to chat with other readers and the author.

What Is Not in This Book

This book does not cover Consul features unrelated to its service mesh functionality. For example, Consul's key/value store and Domain Name System (DNS) service discovery are not covered. Also, this book is not a detailed production-ready operations guide to Consul. The aim is to familiarize readers with Consul's concepts and get them "up and running" with its functionality.

Conventions Used in This Book

The following typographical conventions are used in this book:

Italic
> Indicates new terms, URLs, email addresses, filenames, and file extensions.

`Constant width`
> Used for program listings, as well as within paragraphs to refer to program elements such as variable or function names, databases, data types, environment variables, statements, and keywords.

`Constant width bold`
> Shows commands or other text that should be typed literally by the user. Also used occasionally in program listings to highlight text of interest.

This element signifies a tip or suggestion.

This element signifies a general note.

This element indicates a warning or caution.

Using Code Examples

Supplemental material (code examples, exercises, etc.) is available for download at *https://oreil.ly/consul-examples*.

If you have a technical question or a problem using the code examples, please send email to *bookquestions@oreilly.com*.

This book is here to help you get your job done. In general, if example code is offered with this book, you may use it in your programs and documentation. You do not need to contact us for permission unless you're reproducing a significant portion of the code. For example, writing a program that uses several chunks of code from this book does not require permission. Selling or distributing examples from O'Reilly books does require permission. Answering a question by citing this book and quoting example code does not require permission. Incorporating a significant amount of example code from this book into your product's documentation does require permission.

We appreciate, but generally do not require, attribution. An attribution usually includes the title, author, publisher, and ISBN. For example: "*Consul: Up and Running* by Luke Kysow (O'Reilly). Copyright 2022 Luke Kysow, 978-1-098-10614-0."

If you feel your use of code examples falls outside fair use or the permission given above, feel free to contact us at *permissions@oreilly.com*.

O'Reilly Online Learning

 For more than 40 years, *O'Reilly Media* has provided technology and business training, knowledge, and insight to help companies succeed.

Our unique network of experts and innovators share their knowledge and expertise through books, articles, and our online learning platform. O'Reilly's online learning platform gives you on-demand access to live training courses, in-depth learning paths, interactive coding environments, and a vast collection of text and video from O'Reilly and 200+ other publishers. For more information, visit *https://oreilly.com*.

How to Contact Us

Please address comments and questions concerning this book to the publisher:

O'Reilly Media, Inc.
1005 Gravenstein Highway North
Sebastopol, CA 95472

800-998-9938 (in the United States or Canada)
707-829-0515 (international or local)
707-829-0104 (fax)

We have a web page for this book, where we list errata, examples, and any additional information. You can access this page at *https://oreil.ly/consul-up-and-running*.

Email *bookquestions@oreilly.com* to comment or ask technical questions about this book.

For news and information about our books and courses, visit *https://oreilly.com*.

Find us on LinkedIn: *https://linkedin.com/company/oreilly-media*

Follow us on Twitter: *https://twitter.com/oreillymedia*

Watch us on YouTube: *https://youtube.com/oreillymedia*

Acknowledgments

First and foremost, I would like to thank all of the contributors to Consul over the years. Consul's creators Armon Dadgar and Mitchell Hashimoto, current and former HashiCorp employees, and hundreds of community contributors have made Consul the unique software it is today. It is a privilege to work with you all and write about Consul.

Thanks to Paul Banks and Matthew Keeler for their Consul knowledge, Sabeen Syed for her support, and Hannah Hearth for designing the UI for the sample application. Finally, I owe a debt of gratitude to my reviewers, Nitya Dhanushkodi, Brandon McRae, Guy Barros, and Isha, along with my O'Reilly editors, Corbin Collins, Liz Wheeler, and Gregory Hyman. Thank you so much for your insights, suggestions, and encouragement. This book wouldn't be what it is without you.

## CHAPTER 1

Service Mesh 101

To get started on your service mesh journey, you need to know three things: what a service mesh is, how it works, and why you should use it (and when you should not).

There is no universally accepted definition for a service mesh, but I define it as follows:

> A service mesh is an infrastructure layer that enables you to control the network communication of your workloads from a single control plane.

We can break that definition down into parts to better understand it:

By *infrastructure layer*, I mean that a service mesh is not part of your services; it is deployed and operated independently. Since it is not aware of service-specific business logic, but it affects every service, it is considered infrastructure or *middleware*.

Figure 1-1 shows a typical software stack. Services and applications run on top of infrastructure. Service mesh is at the first infrastructure layer with storage, metrics, and other higher-level infrastructure requirements. Under that is VMs, Kubernetes, or any compute provider or orchestrator where everything runs. At the bottom is actual hardware (bare metal).

By *control the network communication of your workloads*, I mean that a service mesh controls the traffic entering and leaving a microservice, database, or anything else that does network communication. For example, a service mesh might disallow incoming traffic based on a rule (such as it's missing a required header), or it might encrypt outgoing traffic. A service mesh has complete control over all traffic entering and leaving the services.

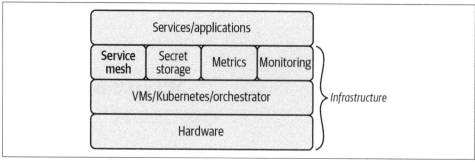

Figure 1-1. A typical software stack

Finally, by *from a single control plane*, I mean a single location from which service mesh operators can interact with the service mesh. Suppose operators want to change the configuration for multiple services. In that case, they don't need to reconfigure a dozen subsystems or modify the services themselves; instead, they configure the service mesh once, and it handles propagating out all changes.

Hopefully, this definition gives you some idea of what a service mesh is, but I often find that I need to understand *how* something actually works before I fully grasp *what* it is.

How a Service Mesh Works

A service mesh is made up of *sidecar proxies* and the *control plane*.

Sidecar Proxies

A *proxy* is an application that traffic is routed through on the way to its destination. Popular proxies you may have heard of are NGINX, HAProxy, and Envoy. In most service meshes, all service traffic (inbound and outbound) is routed through a local proxy dedicated to each service instance.[1]

Figure 1-2 shows what a service mesh looks like with two service instances: frontend and backend. When frontend calls backend, frontend's local proxy captures the outgoing request. frontend's proxy then forwards the request to the backend service. When the request reaches the backend service, again, it is captured by backend's local proxy and inspected. If the request is allowed, backend's proxy forwards it to the actual backend service.

---

1 Some meshes use other technology such as iptables or eBPF to control traffic rather than a separate proxy process.

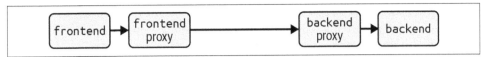

Figure 1-2. Two service instances communicating through the service mesh

Each instance of a service must be deployed with its own local proxy.[2] The pattern of deploying a helper application—in this case a proxy—alongside the main service is known as the *sidecar* pattern, and so local proxies are referred to as *sidecar proxies*.

Sidecar proxies are a vital component of the service mesh because they enable control of service traffic without modifying or redeploying the underlying services. Since sidecar proxies run as separate processes from the services, they can be reconfigured without affecting the services. For example, the backend service's sidecar proxy from Figure 1-2 could be reconfigured to refuse traffic from the frontend service without changing code or redeploying the backend service itself.

What handles reconfiguring proxies? The control plane.

Control Plane

The *control plane's* job is to manage and configure the sidecar proxies. As you can see in Figure 1-3, the control plane is a separate service that must be deployed on its own; it is not deployed as a sidecar. The control plane is where most of the complex logic of the service mesh lives: it must watch for services starting and stopping, sign and distribute certificates, reconfigure proxies, etc. The sidecar proxies themselves are relatively simple: they receive configuration from the control plane detailing which actions to perform on traffic, and they perform those actions.

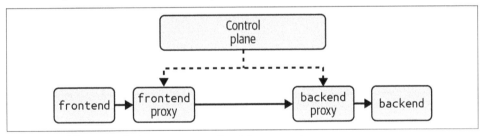

Figure 1-3. The control plane manages the sidecar proxies

If we go back to my definition of a service mesh—"an infrastructure layer that enables you to control the network communication of your workloads from a single control plane"—you can now see how the proxies and control plane fit in.

---

2 If it's impossible to deploy a local proxy—for example, with a managed service such as Amazon Relational Database Service—you can use a terminating gateway as covered in Chapter 10.

The control plane is the single location service mesh operators interact with. In turn, it configures the proxies that control network communication. Together, the control plane and proxies make up the infrastructure layer.

Concrete Example

Let's go through a concrete example to show how a service mesh works in practice. Figure 1-4 shows the architecture for this example.

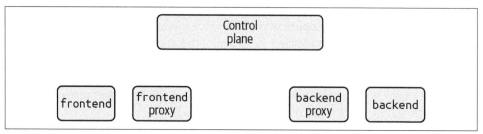

Figure 1-4. Example service mesh architecture

When `frontend` calls `backend`, `frontend`'s sidecar proxy captures the request. In Figure 1-5, the service mesh has configured the `frontend`'s proxy to pass traffic through to the `backend` service without modification. The sidecar proxy running alongside `backend` captures the incoming traffic and forwards the request to the actual `backend` service instance. The `backend` service instance processes the request and sends a response that returns along the same path.

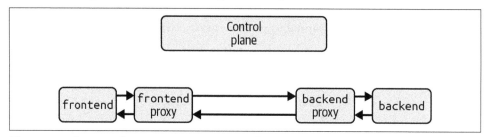

Figure 1-5. frontend's request to backend, and backend's response, is routed through both proxies

Now imagine that you've got a new requirement to get metrics on how many requests per second `frontend` is making to `backend`. You could make changes to the code of `frontend` and `backend` to emit these metrics, but with the service mesh in place, there's a simpler way, as shown in Figure 1-6. First, you configure the control plane with the URL of your metrics database (step 1). Immediately, the control plane reconfigures both sidecar proxies and instructs them to emit metrics (step 2). Now when frontend calls backend (step 3), each proxy emits metrics to the metrics database (step 4), and you can see the requests per second in your metrics dashboard.

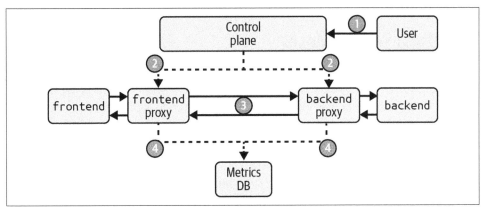

Figure 1-6. The service mesh emitting metrics

Notice that you didn't have to change the code of either service nor did you need to redeploy anything. With a single configuration change, you immediately got metrics for the frontend and backend services.

This concrete example should help you understand how a service mesh works in practice, but it is a simplified picture. In a typical service mesh deployment, the control plane manages hundreds of services and workloads, so the architecture looks more like Figure 1-7. It looks like a mesh, hence the name!

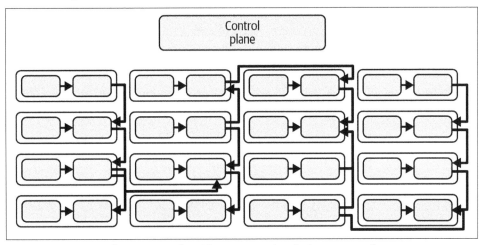

Figure 1-7. A typical service mesh with many services being managed

With a larger mesh, you can see how being able to control networking across all of these services from a single location without redeploying any services or changing code is incredibly powerful. That brings us to *why* you would use a service mesh.

Why Use a Service Mesh

A service mesh provides features in four areas: security, observability, reliability, and traffic control. The fundamental value proposition of a service mesh is the ability to provide these features across *every* service and workload *without modifying service code*.

In the following sections, I will expand upon these areas, but it's important to note that the features a service mesh provides can also be implemented in service code! The question to ask is that if these features can be implemented in service code, why deploy a service mesh at all? The answer is that past a certain scale, recoding every service is more costly to engineering time than running a service mesh. This will be addressed more fully in "When to Use a Service Mesh" on page 11.

Security

One of the primary reasons companies deploy service meshes is to secure their networks. Typically this means encrypting traffic between all workloads and implementing authentication and authorization.

Solving this problem can be very difficult in a microservices architecture without a service mesh. Requiring every request to be encrypted means provisioning Transport Layer Security (TLS) certificates to every service in a secure way and managing your own certificate signing infrastructure. Authenticating and authorizing every request means updating and maintaining authentication code in every service.

A service mesh makes this work much easier because it can issue certificates and configure sidecar proxies to encrypt traffic and perform authorization—all without any changes to the underlying services (see Figure 1-8).

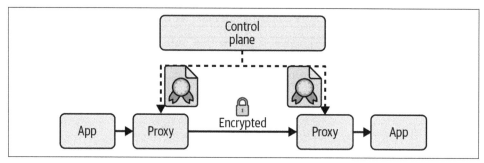

Figure 1-8. A service mesh issuing certificates and encrypting traffic

Security Case Study

Annika is the lead engineer on her platform team. A requirement has come down from the chief information security officer that all traffic between microservices must be TLS encrypted by the end of the year.

Annika knows that modifying every service to support making and receiving requests using TLS is a monumental task. There are 30 development teams that own hundreds of services, some of which haven't been updated for years. Getting space on each team's roadmap to update every service they own will be extremely difficult given all the feature work planned for the year. And even if they could convince the teams to modify and redeploy all their services, they'd still need to build a tool to distribute and rotate the TLS certificates securely.

Luckily, Annika has heard about service meshes. Annika does not need the teams to update any service code with a service mesh. Instead, the sidecar proxies can automatically perform encryption and decryption without the underlying services even being aware. The service mesh also handles securely distributing and rotating TLS certificates. The platform team just needs to deploy the proxies alongside each service; Annika is confident her team can roll that out in a couple of months.

Observability

Observability is the ability to understand what's happening to your services while they're running. Observability data is essential for understanding microservices architectures and diagnosing failure, but it can be challenging to configure all your services to emit metrics and other data in a unified way.

Capturing observability data is the perfect job for a service mesh because all requests flow through its proxies. The service mesh can configure its proxies to emit metrics across all your services in a consistent format without modifying or redeploying the underlying services.

Observability Case Study

Geordi is in a predicament. He just joined the operations team at a startup founded three years ago that's been cranking out features at a breakneck pace. The startup is now thriving and is attracting thousands of new users every day. Unfortunately, all these new users have been putting a ton of load on the system, and the operations team has been experiencing outages more and more frequently.

The worst part is that because the development teams built features so quickly, they never added metrics to their microservices. This means that the operations team is flying blind whenever the site goes down. They have to dig through thousands of logs to find out where the issue is and which service is acting up.

Geordi would love to get all the development teams to update their services to add proper metrics around request response times and errors, but that could take months or even a whole year given how much work the development teams currently have.

Instead, Geordi suggests that the team deploy a service mesh. A service mesh will automatically emit detailed metrics for every request in the system—all without modifying the underlying services. The operations team can then build dashboards for every service and will be able to see which services are returning errors and running slowly. They can even build in alerting thresholds to catch issues before they result in a total outage.

Reliability

In distributed systems, there's often something failing. Building reliable distributed systems means reducing failure where possible and handling failure gracefully when it inevitably happens.

Reducing failure might mean implementing health checking so that traffic is only sent to healthy services. Handling failure might mean retrying requests that failed (see Figure 1-9) or implementing a timeout, so a service doesn't wait forever for a response.

Implementing these techniques in code is time-consuming, error-prone, and difficult to do in a consistent way across all your services. With a service mesh, the proxies can perform these techniques for any of your services—all you need to do is interact with the control plane. You can also adjust the settings in real time as service loads change.

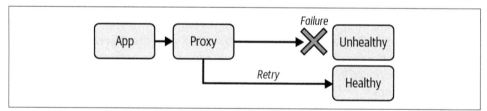

Figure 1-9. The service mesh can be configured to retry failed requests to other instances

Reliability Case Study

Miles just awoke to the sound of the PagerDuty app beeping harshly at him. He's on call, so he rolls groggily out of bed and makes his way to his work bag to pull out his laptop. The alert is that the main home page is down. That's not good.

He loads the home page himself, and his browser just spins. He looks at the monitoring dashboards and sees that latency is off the charts for requests to the home-page service. It's taking five minutes for each request! The weird thing is that the CPU load for the nodes that host the home page looks normal. He wonders if perhaps a dependency of the home page is running slow?

His company is running a service mesh, so he opens up its UI and checks the status of the home-page service. He immediately notices that the analytics service is listed as a dependency of the home-page service and that it's registering an average latency of five minutes. That might explain the outage! The home-page service logs every request to the analytics service. If the home-page service hasn't implemented a timeout on its request to the analytics service, then it will just sit and wait until that request completes.

The frustrating thing is that the request to the analytics service isn't even that important. The company would much rather the site be up and some analytics not recorded than the whole site be down!

Without a service mesh, Miles would need to wake up the home-page developers and get them to implement a timeout in the code and then do a full redeploy. Luckily, the service mesh lets Miles control traffic for any service from the control plane. Miles pushes a new configuration that sets the timeout for requests to the analytics service to 100 milliseconds. He then tries again to load the home page: it loads almost instantly. Miles shoots a message to the on-call chat channel that they need to look at what's going wrong with the analytics service in the morning. Then he goes back to bed.

Traffic Control

Traffic control is about controlling where traffic between services is routed. Traffic control solves many problems:

- Implementing deployment strategies such as *canary deployments*, in which a small amount of "canary" traffic is routed to the new version of a service to see if it's working before fully rolling out the new version
- Monolith to microservices migrations, in which services are split off from the monolith and traffic previously routed to the monolith is seamlessly rerouted to the new microservices
- *Multi-cluster failover*, in which traffic is routed to services in other healthy clusters if the local cluster is down

Traffic Control Case Study

B'Elanna is excited. She's finally got the go-ahead to split out some of the large monolithic service she works on into microservices. The rest of her organization already uses microservices, but her service is the oldest and has grown large and bloated.

Her team plans to split out two microservices from the monolith, a member service and a cart service, and they need to do so without any downtime. Many other microservices depend on the monolith, so as they split it up, they need to ensure that those microservices call the newly split-out services.

Without a service mesh, B'Elanna would need to update all the dependent services to call the new member and cart services instead of the monolith. For example, requests to the /members endpoint now need to go to the new member service, and requests to the /cart endpoint need to go to the new cart service. This means updating a lot of code across many services.

Luckily, B'Elanna's company is running a service mesh. As shown in Figure 1-10, she can configure the service mesh to automatically route any request matching /members to the new member service and likewise for the cart service. The dependent services can continue making their requests as normal and the service mesh handles all the routing. This means B'Elanna can focus on splitting up the monolith rather than updating all the dependent services.

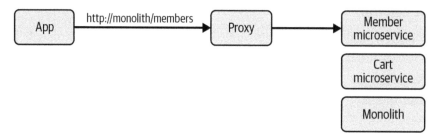

Figure 1-10. The service mesh can be configured to route certain requests to different services

Features in Combination

Now you should understand the types of features a service mesh provides around security, observability, reliability, and traffic control. Alone, these features are helpful, but they are even more powerful in combination.

For example, observability data provided by the service mesh can be combined with reliability and traffic control features. If the mesh detects that a service instance is returning errors, it can redirect traffic to healthy instances or a whole different cluster. Or mesh security features can be combined with the observability features to detect when a service is attempting to make requests it's not authorized to make—potentially indicating a security breach. When you deploy Consul yourself, you'll see many use cases where you can combine the service mesh features.

If your organization needs these features, you must decide whether it's worth the additional complexity of a service mesh, or if you should implement them in service code. The key to answering that question is examining your scale.

When to Use a Service Mesh

There is no doubt that deploying a service mesh adds additional complexity. You now have sidecar proxies and the service mesh control plane to manage. In addition, you will need more compute resources (CPU and memory) to run the proxies and control plane, and now all traffic takes an extra hop through the local sidecar proxies, which adds latency. Implementing service mesh features in code would save resources and reduce infrastructure complexity (although it would add code complexity). For a service mesh to be worth it, it must provide a lot of value to your organization.

A simple formula for knowing when to use a service mesh is when you (a) need to solve networking problems in the areas outlined previously (security, observability, reliability, and traffic control) and (b) your organization is at a scale, or will soon be at a scale, where it's too costly to solve those problems in service code.

For example, say your organization is moving to what's known as a zero trust security architecture where all internal traffic is encrypted, authenticated, and authorized. If you're only running two microservices, you can easily recode those services. However, if you're running 400 microservices, then it's unlikely that you'll be able to recode all those services in a reasonable amount of time. In this case, a service mesh makes a lot of sense.

In addition, at a certain scale, there will be services and workloads that you want control over where you don't actually have the ability to edit their code. For example, maybe you're deploying a packaged open source software, or perhaps you're using a cloud-managed database. Ideally, you would have the same control over those workloads that you do over your other services.

In the end, the exact scale at which it makes sense to use a service mesh will depend on your specific organization and the problems you're trying to solve. I hope that this book will help you understand the problems a service mesh solves and help you gauge whether it makes sense in your situation.

Shared Libraries?

Many engineers who hear the question "How would I implement this across every service?" immediately think about shared libraries. A shared library is a good solution for smaller organizations, but it has many problems past a certain scale.

First, someone has to write the shared library or choose a good one from the open source options. If you use multiple programming languages, your library must support all of them (and any future languages). Second, you must recode all your services to use that library and redeploy them. In large organizations, this is a massive undertaking. Third, if you ever need to update that library (and you will), you then need to update and redeploy every service again.

In some niche use cases, shared libraries are a good option, but they are a poor solution for solving the problems a service mesh solves.

Summary

In this chapter, you learned what a service mesh is, how it works, and why you'd use one.

I introduced my definition of a service mesh:

> An infrastructure layer that enables you to control the network communication of your workloads from a single control plane.

And I discussed how the two components of a service mesh, the proxies and the control plane, enable the control of network communication. You walked through a real-life example of a working service mesh, and I discussed the four categories of service mesh features: security, observability, reliability, and traffic control.

Finally, I addressed when you should use a service mesh: when you need these features and you're at a scale where it's too costly to implement them in service code.

So far, everything discussed in this chapter has been applicable to most service meshes and not specific to Consul. The next chapter is devoted to Consul in particular. You'll learn about how it works, its architecture, the protocols it uses, and what makes it unique.

## CHAPTER 2

Introduction to Consul

> Today we announce Consul, a solution for service discovery and configuration. Consul is completely distributed, highly available, and scales to thousands of nodes and services across multiple datacenters.
>
> —Armon Dadgar (HashiCorp cofounder), April 2014

I remember when Consul first burst onto the scene in 2014. Cloud computing and service-oriented architectures (the precursor to microservices) were becoming mainstream, and every company was starting to grapple with the problem of how to route to services and handle failure in a distributed system.

Consul was a revolutionary technology because it combined DNS-based service discovery with a robust failure detection system. A service would register into Consul and other services could use its Consul DNS entry to route to it. For example, the `frontend` service would be available at `frontend.service.consul`. Consul also detected failure by using a gossip algorithm called Serf (covered later in this chapter) and health checking. If a node or service went down, Consul would quickly notice and remove it from DNS.

Consul was open source, free to use, and it elegantly solved a problem experienced by thousands of companies. The industry quickly adopted it.

Over time, service-oriented architectures grew larger and turned into microservices architectures with smaller and smaller services. This led to the rise of Docker and container orchestrators like Kubernetes. The microservices movement and Kubernetes changed the industry in two important ways.

First, DNS service discovery was no longer sufficient. Developers needed more networking features around security, observability, reliability, and traffic control to help run all these services. It was also becoming more challenging to implement these features in service code because the number of services kept growing.

Second, Kubernetes made it much easier to run sidecar proxies thanks to the pod model, where multiple containers can run together in a private network.[1]

These changes enabled service mesh technology as we know it today.

The Consul team at HashiCorp was following these trends, and in 2018 they released a service mesh feature called Consul Connect. Consul Connect focused on secure communication between services and added support for sidecar proxies to encrypt traffic between services. Since then, Consul has developed into a fully functional service mesh with features not just around secure communication but also observability, reliability, and traffic control.

With that history out of the way, you're ready to learn about how Consul works and what makes it unique as a service mesh. The following sections cover Consul's architecture and the protocols it uses to remain reliable and scalable.

You will see mention of Consul Connect in various configuration settings. This refers to Consul's service mesh features.

Architecture

A Consul installation is made up of three sets of components: Consul servers, Consul clients, and sidecar proxies. Sidecar proxies talk to their local Consul clients, and Consul clients talk to Consul servers. There are usually three or five Consul server nodes, and there can be up to thousands of workload nodes (see Figure 2-1).

To relate Consul's architecture back to the architecture of a generic service mesh, as shown in Figure 2-2, Consul servers and clients are part of the control plane.

---

1 Previously, it was possible to run two processes together; however, it was more difficult because you'd need to manually orchestrate their deployment and configuration. In addition, other processes on the same machine could access the same network and filesystem.

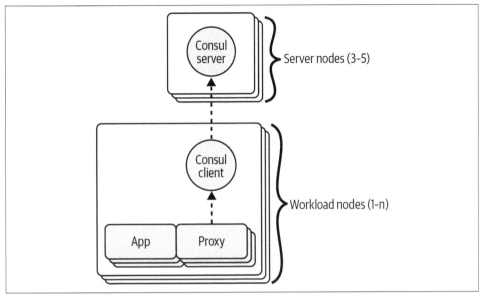

Figure 2-1. A Consul installation consists of Consul servers, Consul clients, and sidecar proxies

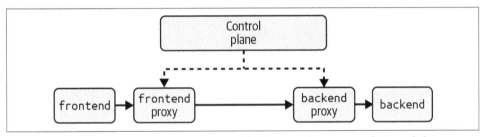

Figure 2-2. Consul's control plane consists of both Consul servers and Consul clients

Consul Servers

Consul servers are Consul's database. Consul needs to store data such as the *catalog* (the list of nodes and services), configuration, health statuses, and more. For a production Consul deployment, you must run multiple Consul servers (although for testing purposes, one is sufficient). Running multiple servers ensures high availability and data persistence: if one server goes down, the remaining servers can still service requests and no data is lost.

You should always run an odd number of servers—for example, three or five—because running an even number of servers doesn't increase Consul's fault tolerance (and so the extra server is a waste of resources). Consul must have a *majority* of servers running to continue functioning without issue.[2] If you're running three servers, a majority is two, so you can tolerate one server being down. If you're running four servers, a majority is three, so you can still only tolerate one server being down.

 Running multiple servers ensures reliability, but it makes it difficult to keep data consistent: what if there's a network partition between the servers and they get out of sync? Consul uses a protocol called *Raft* (see the "Building Consensus with Raft" sidebar) to solve this problem.

Consul servers store their data on disk. On Kubernetes, Consul servers are deployed as a StatefulSet and use PersistentVolumes to store their data:

StatefulSet
A Kubernetes resource type similar to a *Deployment* (a Kubernetes resource type used to deploy multiple replicas of a service and manage its lifecycle) that guarantees the same storage volume is always paired with the same pod.

PersistentVolume
A chunk of storage that can be mounted into a pod but won't be deleted if that pod restarts.

On VMs, Consul servers should be deployed with disk volumes that won't be lost when the node restarts.

For production, it's recommended to dedicate an entire node to each Consul server to avoid resource contention issues. On Kubernetes, this can be accomplished by having a separate node pool that uses taints and tolerations to restrict what pods can be scheduled to those nodes.

Building Consensus with Raft

Consul servers are constantly communicating with one another using Raft. Raft is a *consensus algorithm*: a family of algorithms used in distributed systems to ensure that multiple servers can agree on the state of the world. Raft performs two duties in Consul: leader election and replication.

Leader election is the process of choosing one Consul server as the leader. All writes to Consul—for example, registering a new service—go through the leader. Without a single leader, two different values could be written simultaneously to separate servers

---

2 If a majority of servers aren't running, Consul will "fail static." This means things will continue to work, but no changes (such as registering new services) can be made.

and then it would be unclear what the actual value should be. For example, what IP should be used if the same service was registered with different IP addresses simultaneously on separate servers? With Raft, all registrations go through the leader, and the leader alone decides the true IP for the service. Raft also ensures that if the leader goes down, a new one is elected.

Replication is the process of ensuring all writes to the leader are copied to other Consul servers. If the leader goes down and a new leader is elected, Raft replication guarantees that the new leader will have the same data as the old leader, so nothing is lost.

Consul Clients

Consul clients run on every workload node in the cluster.[3] Consul clients are responsible for detecting the health of services running on their node and the health of other nodes in the cluster. Consul clients then send this information to the Consul servers to keep their catalog up to date.

In theory, Consul servers *could* check the health of all services and nodes in a cluster, and there would be no need for Consul clients. In practice, however, this centralized approach is not scalable to the tens of thousands of nodes and services that Consul aims to support. Instead, running a Consul client on each node allows Consul to use a distributed approach (see "Distributed Failure Detection Using Serf" on page 18) to checking service and node health that easily scales and results in failure detection that takes milliseconds, even in massive clusters.[4]

Consul clients are also responsible for configuring the sidecar proxies on their nodes. This is covered in more detail when I walk through an example use case.

 It's possible for Consul servers to also act as Consul clients. This means you can run services on the same nodes as Consul servers and the servers will manage the sidecar proxies without the need to also run a Consul client. This architecture is only recommended for small test clusters where it doesn't make sense to use separate nodes for Consul servers and service workloads.

---

3 If you have dedicated nodes for Consul servers, then the other nodes in your cluster that run actual services and applications are *workload nodes*.

4 On Kubernetes, Consul's failure detection isn't as necessary because Kubernetes performs its own node and service failure detection via the *kubelet*. In future versions of Consul, Consul clients may no longer be needed on Kubernetes.

Distributed Failure Detection Using Serf

Consul needs to detect two types of failures: service failure and node failure. Service failure is when a service running on a node fails. To detect this, Consul clients run health checks against each service on their node.[5] This approach is scalable because there are usually 1–100 services per node.

Node failures are when entire nodes go offline. If a node goes offline, the Consul client on that node can't tell the rest of the cluster that its services are no longer running, since it is dead! Consul servers could perform similar health checks against nodes just like Consul clients do with services, but this would not be scalable because there could be thousands of nodes and only three or five Consul servers.

Instead, to detect node failure Consul uses a library called Serf. Serf implements a *gossip* algorithm (also known as an epidemic algorithm): a family of algorithms that shares data through a cluster by relying on each client to relay messages to a small number of other clients (by default three), chosen at random. Those clients in turn choose three more random clients to communicate with, and so on and so forth (see Figure 2-3).[6] This is similar to how gossip is spread, or viruses are transmitted, hence the name.

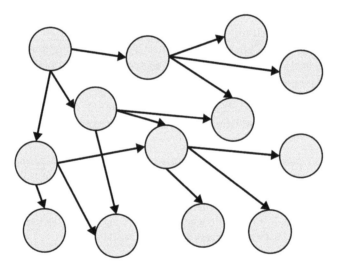

Figure 2-3. A gossip algorithm works by clients communicating with a small number of other clients. Those clients in turn communicate with other clients, and information spreads quickly.

---

5 On Kubernetes, the readiness probe status is synced into Consul. This is covered in more detail in Chapter 8.

6 Clients will choose a random set of nodes each time they relay a message.

In Serf, clients are constantly checking on each other. If a client is determined to be down, that information is quickly gossiped to the rest of the cluster. In theory, distributed failure detection takes longer than a centralized approach, but in practice, failure is detected and gossiped to all clients in less than two seconds, even on thousand node clusters.

On Kubernetes, Consul clients are run as a DaemonSet. A *DaemonSet* is a Kubernetes resource type similar to a Deployment that guarantees one Consul client pod will run on every node in the cluster. This ensures there's always a Consul client available to manage the sidecar proxies. On VMs, it's up to you to ensure that every workload node has a Consul client provisioned.

Sidecar Proxies

Each *service instance*, a specific running instance of a service, has a dedicated sidecar proxy. The proxy's responsibility is to intercept requests to and from its service instance and perform actions upon the requests as configured—for example, encrypt requests with TLS, or refuse requests from certain services.

Consul itself isn't a proxy.[7] Instead, Consul uses a proxy called Envoy (*https://www.envoyproxy.io*). Envoy is an open source proxy created at Lyft. It is now a CNCF (Cloud Native Computing Foundation) project used by many service meshes. Envoy is written in C++ for performance and has a very low memory footprint. This means it can run beside all your services with minimal resource and latency impact. Envoy supports hundreds of features for load balancing, observability, health checking, and more. Consul clients configure Envoy via its gRPC API.

On Kubernetes, sidecar proxies run as separate containers within service pods. This ensures that the service containers can route to their sidecar proxies within the local pod network. On VMs, sidecar proxies run as separate processes for each service instance on that VM.

Now that you're familiar with Consul servers, clients, and sidecar proxies, let's work through an example use case to understand how all three components work together.

Example Use Case

Imagine a new service called backend is started on a node (see step 1 in Figure 2-4). Next, the backend service is registered with its IP address—for example, 1.2.3.4— onto the Consul client on that node (step 2).[8] The Consul client then makes a request

---

7 OK, technically it is because it ships with what's called the *built-in proxy*, but this proxy has minimal functionality and should only be used for small test cases.

8 The exact mechanism for how the service is registered is covered in Chapter 4.

back to the Consul servers to inform them that a new service called backend is running on that node with address `1.2.3.4` (step 3). The Consul servers then add that service to the catalog (step 4).

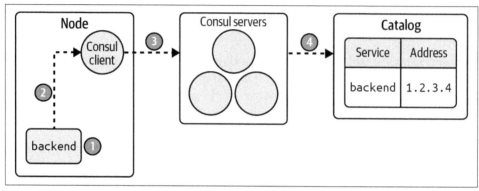

Figure 2-4. A new service being started on a node

At the same time, the proxy for the backend service starts up. The proxy opens a connection to the local Consul client (see step 1 in Figure 2-5). The Consul client configures the proxy and keeps the connection open for future reconfiguration (step 2).

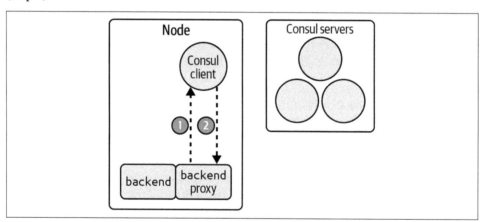

Figure 2-5. Consul client configuring the backend proxy

Now imagine that there's another service called frontend running on a different node in the cluster and that the frontend service is configured to talk to the backend service (shown in Figure 2-6).

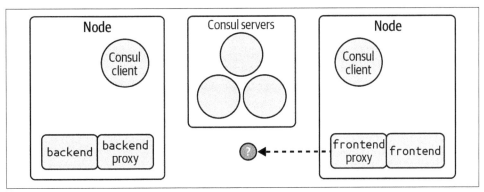

Figure 2-6. The frontend service is configured to talk to the backend service, but it doesn't yet know which address to use

frontend's proxy needs to know the address of the backend service to make a request. To accomplish this, the Consul client on the same node as the frontend service watches the Consul server catalog for new instances of the backend service (step 1 in Figure 2-7). When a new instance of the backend service is registered into the catalog, the Consul servers send the new address to the Consul client (step 2). The Consul client then updates the configuration of the frontend service's proxy with the new address for the backend service (step 3).

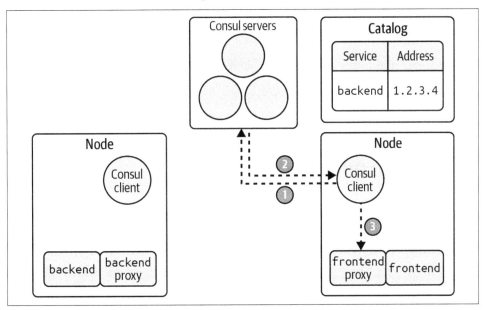

Figure 2-7. frontend's proxy learning the address of the backend service in order to make a request

Now the `frontend` service's proxy knows the address of the `backend` service. When the `frontend` service makes its request, `frontend`'s proxy intercepts it (step 1 in Figure 2-8). The proxy sees that the request is bound for the `backend` service, and since it knows the address of the `backend` service, `1.2.3.4`, it forwards the request to that address (step 2). The request reaches the `backend` service, where the `backend` service's proxy intercepts it (step 3). Assume that `backend`'s proxy is configured to only allow requests from the `frontend` service. The proxy inspects the request and sees that it's from the `frontend` service,[9] and so it allows the request through (step 4).

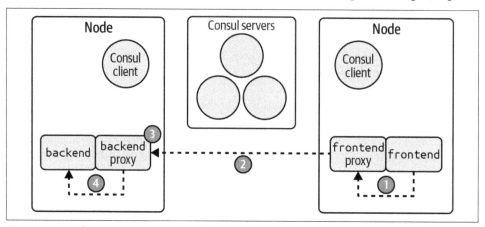

Figure 2-8. Sidecar proxies capture inbound and outbound traffic and follow their configured rules to act on that traffic

This example illustrates how Consul servers, clients, and sidecar proxies work together. Consul servers and clients communicate to share data about the cluster and configure the sidecar proxies. Sidecar proxies capture inbound and outbound traffic and follow their configured rules to act on that traffic, for example by routing it to another proxy or disallowing the traffic because it's not authorized.

Consul Versus Other Meshes

There are many other service meshes in the marketplace, such as Istio (*https://istio.io*) and Linkerd (*https://linkerd.io*). Most services meshes follow the same general architecture with a control plane that manages sidecar proxies. Consul is unique among other meshes in that its control plane can run entirely separate from Kubernetes. This means if you're managing a VM cluster that you don't need to also run Kubernetes.

---

9 The mechanism for verifying the source of requests is covered in Chapter 6.

Each service mesh has its pros and cons depending on your exact use case. A good approach to choosing a service mesh is to focus on your top three use cases, such as security, observability, and multi-cluster support, and then test out the most popular service meshes and choose the one you feel the most comfortable with.

Consul's Other Features

Consul is not just a service mesh. It's also a key/value store for service configuration and a DNS service discovery solution. When using DNS, there are no sidecar proxies. Requests are routed directly between services. This means there is no automatic encryption, observability, or traffic control features.

This book focuses on Consul's service mesh features, but you can visit Consul's documentation (*https://oreil.ly/LShQE*) to learn more about its other features.

Summary

Now you understand how Consul works at a high level. The Consul servers are Consul's database and manage the catalog—a list of all the nodes and services in the cluster. The Consul clients are deployed on every workload node and manage the sidecar proxies running on their local node. Sidecar proxies capture traffic and perform service mesh actions such as encryption and emitting metrics.

You also learned how Consul operates as a reliable and scalable system through the use of Raft for consensus on Consul servers and Serf for failure detection among clients.

In the next chapter, you'll get to actually deploy Consul on Kubernetes or VMs.

## CHAPTER 3

Deploying Consul

In this chapter, you will learn how to deploy Consul on Kubernetes or VMs. You can deploy Consul on self-managed or cloud-managed Kubernetes clusters and VMs. For testing purposes, and so you can complete the exercises throughout this book, this chapter also covers provisioning a Kubernetes cluster or VM on your local machine.

Once Consul is installed, you'll learn how to interact with it through its UI, CLI, and API.

If you're deploying Consul on VMs, skip ahead to "Deploying Consul on VMs" on page 34. Otherwise, continue reading to deploy Consul on Kubernetes.

> If you're unsure if you should use VMs or Kubernetes, I recommend deploying on Kubernetes because there are fewer steps.

Deploying Consul on Kubernetes

To deploy Consul on Kubernetes, you must first have a running Kubernetes cluster. Then you'll use the consul-k8s CLI to install Consul with a single command.

Provisioning a Kubernetes Cluster

Consul can run on any type of Kubernetes cluster, whether in the cloud, for example, on Amazon Elastic Kubernetes Service (EKS) or Google Kubernetes Engine (GKE), or in a self-managed datacenter.

For the exercises in this book, if you're running macOS or Linux, I recommend running Kubernetes on your local machine using a tool called minikube. I recommend minikube because it's free and because it has an automatic port-forwarding feature that makes it easy to access services without fiddling with a lot of kubectl port-forward commands.

If minikube isn't an option for you and you're interested in running Kubernetes in the cloud, see the sidebar "Cloud-Managed Kubernetes". On Windows I found minikube to be unstable, so I recommend using cloud-managed Kubernetes for Windows users.

Cloud-Managed Kubernetes

Running Kubernetes in a cloud provider–managed cluster costs money (unless you're using free credits), but you won't have to worry about installing Docker on your workstation and experiencing excess resource consumption.

See the following links for how to spin up a managed Kubernetes cluster in your favorite cloud provider:

- Google Cloud Platform Google Kubernetes Engine (GKE) (*https://oreil.ly/XV5Ip*)
- Amazon Elastic Kubernetes Service (EKS) (*https://oreil.ly/ZW2Zf*)
- Azure Kubernetes Service (AKS) (*https://oreil.ly/yk5un*)
- Alibaba Cloud Container Service for Kubernetes (ACK) (*https://oreil.ly/SL9F8*)
- DigitalOcean Managed Kubernetes (*https://oreil.ly/89NsK*)

Many of the examples in this book use an automatic port-forwarding feature of minikube to allow access to various workloads over localhost. If you're running your cluster in the cloud, you'll need to run kubectl port-forward yourself. I try to note the port-forward commands in the relevant sections.

Before you provision a Kubernetes cluster (and regardless if you're using minikube or a cloud-managed cluster), you'll need the Kubernetes CLI tool kubectl (*https://oreil.ly/qRjVd*) installed. The exercises in this book were tested against kubectl version 1.23.3.

Follow these install instructions depending on your operating system:

- *Windows: https://oreil.ly/CRJ4Z*
- *macOS: https://oreil.ly/hvB8x*
- *Linux: https://oreil.ly/LLEy1*

On Windows, I recommend using the Chocolatey package manager (*https://oreil.ly/Wtv2L*) to make installing tools easier.

To verify `kubectl` is installed successfully, run:

```
$ kubectl version --client

Client Version: version.Info{Major:"1", Minor:"23", ...
```

Next, it's time to provision a Kubernetes cluster using minikube. If you've decided to use a non-minikube Kubernetes cluster, skip ahead to "Installing Consul with the consul-k8s CLI" on page 29.

Installing minikube

Minikube requires a driver on which to run Kubernetes. I recommend using the Docker driver because it is the most stable, in my experience. Refer to the following instructions for installing the Docker driver:

- *macOS:* "Install Docker Desktop on Mac" (*https://oreil.ly/X8x3v*)
- *Linux:* "Install Docker Engine" (*https://oreil.ly/36GwO*)
- *Windows:* As mentioned, minikube on Windows is unstable, so I recommend using a cloud-managed Kubernetes cluster.

 On macOS you can change the default CPU and RAM allocation for Docker through its preferences panel. I recommend allocating six CPUs and 8 GB of RAM if possible. A lower allocation will work as well, but Consul will run slower.

Next, install minikube v1.25.2:[1]

- macOS (Intel):

```
$ export DOMAIN="https://storage.googleapis.com"
$ curl -LO \
    "$DOMAIN/minikube/releases/v1.25.2/minikube-darwin-amd64"
$ sudo install minikube-darwin-amd64 /usr/local/bin/minikube
$ rm minikube-darwin-amd64
```

---

1 Other minikube versions may work, but the exercises in the book were only tested against v1.25.2.

- macOS (Apple Silicon):

```
$ export DOMAIN="https://storage.googleapis.com"
$ curl -LO \
    "$DOMAIN/minikube/releases/v1.25.2/minikube-darwin-arm64"
$ sudo install minikube-darwin-arm64 /usr/local/bin/minikube
$ rm minikube-darwin-arm64
```

- For Linux and other install options, see minikube's installation guide (*https://oreil.ly/aeQ0E*).

Now you're ready to start minikube. This command should take two or three minutes to complete, depending on your internet connection:

```
$ minikube start --driver=docker --kubernetes-version=v1.23.3
```

When the minikube start command completes, check the status of your cluster:

```
$ minikube status

minikube
type: Control Plane
host: Running
kubelet: Running
apiserver: Running
kubeconfig: Configured
```

minikube should have configured your kubectl context to point at the minikube cluster.[2] Run kubectl get nodes to verify:

```
$ kubectl get nodes
```

NAME	STATUS	ROLES	AGE	VERSION
minikube	Ready	control-plane,master	1m	v1.23.3

> Use kubectl config use-context minikube to switch your context to the minikube cluster if it isn't already.

Now you're ready to install Consul.

---

2 kubectl can be used to manage multiple Kubernetes clusters. It uses *contexts* to determine which cluster is currently being managed.

If you want to temporarily stop your minikube cluster because you're not using it, run `minikube stop`. This will stop minikube and save its state so you don't lose your work. You can then run `minikube start` to restart the same cluster and continue your work.

If you need to fully re-create your cluster from scratch for some reason, you can delete it by running `minikube delete` and then re-create it with the same `minikube start` command you used previously. You'll then need to reinstall Consul.

Installing Consul with the consul-k8s CLI

Now that you've got a Kubernetes cluster ready to go, it's time to install Consul. Consul on Kubernetes can be installed via the consul-k8s CLI or Helm (*https://helm.sh*), a package manager for Kubernetes. I recommend using the CLI because it performs extra checks to ensure the installation will be successful.

To learn how to install Consul using its Helm chart, see the Consul documentation (*https://oreil.ly/gw3Jj*).

Installing the consul-k8s CLI

To install the consul-k8s CLI, navigate to the consul-k8s v0.44.0 binaries web page (*https://oreil.ly/DnrHm*) and click on the link corresponding to your operating system and architecture:

- *On macOS:* Use the darwin zip files: `darwin_amd64.zip` for Intel processors and `darwin_arm64.zip` for Apple Silicon.
- *On Linux:* Run `uname -sm` to see your architecture. If it says `x86_64`, use the `_amd64` package and if it says `i386`, use the `_386` package.
- *On Windows:* Follow the instructions provided on Microsoft's Windows support FAQ page (*https://oreil.ly/lL4z0*) to determine if your system is 32 or 64 bit. If 32, use the `_386` binary; otherwise, the `_amd64` binary.

After you've downloaded the zip file, open it to extract the `consul-k8s` binary. On macOS or Linux, move it to your */usr/local/bin* directory:

```
$ sudo mv <path-to-downloads>/consul-k8s /usr/local/bin
```

On Windows, follow the instructions provided in the Stack Overflow thread "'Register' an .exe So You Can Run It from Any Command Line in Windows" (*https://oreil.ly/HMWgj*) to add `consul-k8s` to your path.

Test that `consul-k8s` is installed:

```
$ consul-k8s version

consul-k8s v0.44.0
```

values.yaml file

Your Consul on Kubernetes installation is configured using a *values.yaml* file. This file sets the specific options for your installation of Consul.[3] For example, this file is used to control how many Consul servers should be installed.

First, choose a directory in which to work—for example, *~/code/consul*:

```
$ mkdir -p ~/code/consul
$ cd ~/code/consul
```

Next, create your *values.yaml* file as shown in Example 3-1.

 All code examples can be found in the book's GitHub repo (*https://oreil.ly/consul-examples*) for easier copying and pasting.

Example 3-1. values.yaml

```
# Settings under "global" pertain to all components
# of the Consul installation.
global:
  # The name of your installation. This should always
  # be set to consul.
  name: consul
  # Enable metrics so you can observe what's happening
  # in your installation.
  metrics:
    enabled: true
  # Consul image.
  image: hashicorp/consul:1.11.5
  # Envoy image.
  imageEnvoy: envoyproxy/envoy:v1.20.2

# Settings under "server" configure Consul servers.
server:
  # replicas sets the number of servers.
  # In production, this should be 3 or 5, however
  # for testing, this should be set to 1.
  replicas: 1
```

---

3 This is the same file used for the Helm chart installation. Under the hood the CLI uses Helm.

```
# Enable Consul's service mesh functionality.
connectInject:
  enabled: true

# Settings under "controller" configure Consul's controller
# that manages custom resources.
# Custom resources are covered in later chapters.
controller:
  enabled: true

# Install Prometheus, a metrics database.
prometheus:
  enabled: true

# Settings under "ui" configure the Consul UI.
ui:
  service:
    # Use a load balancer service in
    # front of the Consul UI so we can access it using
    # minikube tunnel.
    type: LoadBalancer
    # Use port 8500 for the UI.
    port:
      http: 8500
```

 If running a Kubernetes cluster in the cloud or on Linux with minikube, set the UI's service type to ClusterIP instead of Load Balancer.

In the cloud, leaving it as LoadBalancer might expose the UI to the public internet, which is insecure. On Linux, there's no reason to set it to LoadBalancer because you won't be using minikube tunnel.

In both these cases, you will access the UI using kubectl port-forward.

Install

To install Consul, you will use the consul-k8s install command.

Before you do, if you're using minikube on macOS you must start a minikube tunnel:

```
$ minikube tunnel

Tunnel successfully started
```

Leave this command running as you progress through the book. If you need to stop and restart your minikube cluster, be sure to also restart minikube tunnel.

The minikube tunnel command ensures that load balancer services are allocated external IPs on minikube. The consul-k8s install command will wait until all services have IPs before exiting, so you must have minikube tunnel running for the installation to complete. On cloud-managed and Linux minikube clusters, you've set the UI service to ClusterIP so you don't need minikube tunnel.

Keep minikube tunnel running (if applicable) and, in another terminal, navigate to your *values.yaml* directory and run consul-k8s install:

```
$ consul-k8s install -config-file values.yaml

...
Consul installed in namespace "consul".
```

It can take up to 10 minutes for the installation to complete, but it usually takes about 3 minutes.

If you're using minikube and it's complaining that Service does not have load balancer ingress IP address, ensure you're running minikube tunnel.

Verifying your installation

To verify your installation, run consul-k8s status (it may take a minute for the Consul clients to become healthy):

```
$ consul-k8s status

...
Consul servers healthy (1/1)
Consul clients healthy (1/1)
```

The Appendix lists issues you may encounter during the course of completing the exercises throughout the book and how to fix them.

You can look at the resources deployed to your cluster using kubectl:

```
$ kubectl get daemonset,statefulset,deployment -n consul

NAME
daemonset.apps/consul-client

NAME
statefulset.apps/consul-server

NAME
deployment.apps/consul-connect-injector
deployment.apps/consul-controller
deployment.apps/consul-webhook-cert-manager
deployment.apps/prometheus-server
```

As described in Chapter 2, there is a DaemonSet for the Consul clients and a Stateful-Set for the Consul servers. There are also four deployments that run Kubernetes-specific automation:

consul-connect-injector

> This deployment handles automatically injecting service mesh sidecars and syncing Kubernetes probes to Consul. This functionality is covered in more detail in later sections and chapters.

consul-controller

> This deployment manages Kubernetes custom resources. Custom resources are covered in Chapter 5.

consul-webhook-cert-manager

> This deployment handles creating certificates needed by Consul to communicate with Kubernetes securely.

prometheus-server

> This deployment runs a metrics database called Prometheus that collects metrics from Consul. Prometheus is covered in more detail in Chapter 7.

Congratulations! You've successfully installed Consul on Kubernetes. Now you're ready to interact with Consul via its UI, CLI, and API. Skip ahead to "Interacting with Consul" on page 40, or read on to learn how to deploy Consul on VMs.

> Since there is only one node in the minikube cluster, both the Consul server and client pods are running on the same node. This is not a problem for test clusters, but Consul servers should run on separate nodes for production clusters.

Deploying Consul on VMs

One of Consul's strengths is that it runs just as well on VMs as on Kubernetes. Consul can run on all types of VMs (and bare-metal machines), and it can scale to clusters with thousands of nodes.

In a production cluster, you should dedicate three or five VMs for running Consul servers.[4] The rest of your VMs will run Consul clients and your service workloads.

 The Consul Reference Architecture (*https://oreil.ly/nJRhV*) has recommendations for VM sizing and resource requirements for production clusters.

To test out Consul and follow the exercises throughout this book, I recommend running a single Linux VM on your local machine that I've prepackaged with all the tools you need. This single VM will run one Consul server that will also act as a Consul client.

First, you need to provision the VM, and then you will configure and run Consul using `systemd`.

Provisioning a Local VM

You're going to use a program called Vagrant that makes it easy to manage VMs on a local workstation.[5]

To use Vagrant, install the `vagrant` CLI tool (*https://oreil.ly/sluCe*) and VirtualBox (*https://oreil.ly/DoxLu*), an application for running VMs locally. If running macOS on an Apple Silicon machine, follow the instructions in this GitHub Gist (*https://oreil.ly/2oZKJ*).

Next, choose a directory in which to work—for example, *~/code/consul*—and navigate into it:

```
$ mkdir -p ~/code/consul
$ cd ~/code/consul
```

Create a file *Vagrantfile* that describes how to configure the VM, as shown in Example 3-2.

---

4 In very large clusters with high usage, some companies run even more servers.

5 Vagrant was also created by HashiCorp and was the company's first product.

Example 3-2. Vagrantfile

```
Vagrant.configure("2") do |config|
  config.vm.box = "consul-up/vm" ❶

  config.vm.network "forwarded_port", guest: 3000, host: 3000 ❷
  config.vm.network "forwarded_port", guest: 8500, host: 8500
  config.vm.network "forwarded_port", guest: 6060, host: 6060
  config.vm.network "forwarded_port", guest: 8080, host: 8080
  config.vm.network "forwarded_port", guest: 9090, host: 9090
  config.vm.network "forwarded_port", guest: 16686, host: 16686
end
```

❶ This line references a VM that I've created especially for readers. It has a number of tools preinstalled on it.

❷ These directives make certain ports on the VM accessible on localhost.

> All code examples can be found in the book's GitHub repo (*https:// oreil.ly/consul-examples*) for easier copying and pasting.

Confirm that Vagrant is installed and sees the *Vagrantfile* by running `vagrant status`:

```
$ vagrant status
```

```
Current machine states:
default                  not created (virtualbox)
```

Because you haven't started the VM yet, it should show the state as `not created`.

> You must be in the same directory as your *Vagrantfile* to run `vagrant` commands.

To download and start the VM, run:

```
$ vagrant up
```

This process should take about five minutes (depending on your internet connection). Once the command completes, run `vagrant status` again to check that the VM has started:

```
$ vagrant status
```

```
Current machine states:
default                running (virtualbox)
...
```

 When you're not using the VM, you can save system resources by suspending it using `vagrant suspend`. This command will save the state of the VM. To restart it, navigate to the directory with the *Vagrantfile* and run `vagrant up`.

You can now Secure Shell (SSH) into the VM:

```
$ vagrant ssh
```

Installing and Configuring Consul

Since this is a VM I created for readers, Consul is already installed. To confirm, run `consul version`:

```
$ consul version
```

```
Consul v1.11.5
Revision d8983fc9
Protocol 2 spoken by default...
```

If you are provisioning your own VMs, see Consul's download page (*https://oreil.ly/hKBP0*) for instructions on installing Consul.

This book uses Consul version 1.11.5. I recommend that you use this version to complete the exercises throughout this book, even if newer versions of Consul are available. This will ensure all the commands and code samples in the book work.

With Consul installed, you're ready to configure it. Consul can be configured via environment variables, flags, and configuration files, but some settings can only be configured using configuration files. For this reason, you'll be using configuration files.

There is an existing configuration file at */etc/consul.d/consul.hcl*. This directory and file were created when I installed the Consul package as part of preparing the VM. If you open up */etc/consul.d/consul.hcl*, you will see that it already has some configuration set. All of the configuration values are commented out except for `data_dir`:

```
# data_dir
# This flag provides a data directory for the agent to store state...
data_dir = "/opt/consul"
...
```

Consul supports JSON and HCL configuration files. HCL stands for HashiCorp Configuration Language. It is similar to JSON but is more human-readable and -writable.

The configuration in */etc/consul.d/consul.hcl* is only a starting point. You need to add additional configuration to enable the service mesh features of Consul and to configure Consul to run as a server.

Create a new file */etc/consul.d/server.hcl* and open it with your favorite editor.

 The Consul configuration files are only writable by the consul or root users, so you'll need to use sudo to edit them. For example:

```
$ sudo vim /etc/consul.d/server.hcl
```

Add the contents from Example 3-3 to the file.

Example 3-3. /etc/consul.d/server.hcl

```
# connect
# This stanza configures connect, the name
# for the service mesh features of Consul.
connect {
  enabled = true
}

# ports
# Configures which ports Consul listens on.
# You need to configure its gRPC port to listen on 8502
# because this is required for the service mesh functionality.
ports {
  grpc = 8502
}

# server
# Configures this agent to run as a server (as opposed to a client).
server = true

# bootstrap_expect
# Sets the number of servers expected to be in this cluster.
# Since you only have one server, this is set to 1.
# The servers will wait until this many servers
# have joined the cluster before they start up.
bootstrap_expect = 1

# ui_config
# Configures Consul's UI.
```

```
# Set enabled to true to enable the UI.
ui_config {
  enabled = true
}

# client_addr
# The address Consul binds to for its HTTP API.
# The UI is exposed over the HTTP API so to access
# the UI from outside the VM, set this to 0.0.0.0 so it
# binds to all interfaces.
client_addr = "0.0.0.0"

# bind_addr
# The address Consul binds to for internal cluster
# communication. Usually this should be set to
# 0.0.0.0 but in Vagrant, setting this to 127.0.0.1
# prevents issues if the IP changes.
bind_addr = "127.0.0.1"
```

Consul is now configured and ready to be started. You need to start Consul using the operating system's service manager so it can continue running in the background. On Ubuntu, the Linux operating system used by this VM, the manager is called systemd.

systemd

systemd operates on *unit files* that describe how to run each service. Consul's unit file */usr/lib/systemd/system/consul.service* shown in Example 3-4 was created when the Consul package was installed.

 A *unit* in systemd means any resource managed by systemd. A *unit file* is a systemd configuration file that tells systemd how to run and manage that unit.

Example 3-4. /usr/lib/systemd/system/consul.service

```
[Unit]
Description="HashiCorp Consul - A service mesh solution"
Documentation=https://www.consul.io/
Requires=network-online.target
After=network-online.target
ConditionFileNotEmpty=/etc/consul.d/consul.hcl

[Service]
User=consul
Group=consul
ExecStart=/usr/bin/consul agent -config-dir=/etc/consul.d/
ExecReload=/bin/kill --signal HUP $MAINPID
KillMode=process
```

```
KillSignal=SIGTERM
Restart=on-failure
LimitNOFILE=65536

[Install]
WantedBy=multi-user.target
```

Before you instruct `systemd` to start the `consul` service, enable it to start at boot via the `systemctl enable` command. This ensures if you restart your VM, Consul will also restart:

```
$ sudo systemctl enable consul
```

```
Created symlink /etc/systemd/system/...
```

Now you can start the `consul` service:

```
$ sudo systemctl start consul
```

Check its status:

```
$ sudo systemctl status consul
```

```
consul.service - "HashiCorp Consul - A service mesh solution"
    Loaded: loaded (/lib/systemd/...
    Active: active (running) since ...
    ...
```

If its status is `active (running)`, then Consul is running!

You can view Consul's logs with `journalctl`:

```
$ journalctl -u consul
```

```
systemd[1]: Started "HashiCorp Consul - A service mesh solution".
consul[31615]: ==> Starting Consul agent...
consul[31615]:            Version: '1.11.5'
consul[31615]:            Node ID: '01ec3a1e-11c5-9a04-0a62-65afaf4447b2'
consul[31615]:          Node name: 'vagrant'
consul[31615]:         Datacenter: 'dc1' (Segment: '<all>')
consul[31615]:             Server: true (Bootstrap: true)
consul[31615]:        Client Addr: [0.0.0.0] (HTTP: 8500, HTTPS: -1, ...
...
```

- If the logs take up the whole screen, scroll through them with the j and k keys for down and up, respectively.

- To exit `journalctl` if it's in full-screen mode, type q.

- To tail the logs, use the `-f` flag—`journalctl -u consul -f`—and then use `Ctrl-c` to exit.

Congratulations! You've got a Consul server up and running on your VM. The exercises in this book only require a single Consul node, so there is no more setup required. See the Consul in Production sidebar for more information about the additional steps needed to run Consul in production.

Consul in Production

You'll need more than a single node for your production cluster. As shown in Chapter 2, you'll need to run three or five Consul server nodes and as many workload nodes as necessary for all your services.

The server nodes will be configured similarly to the node in this exercise, but they'll also be configured with the addresses of the other server nodes using the `retry_join` (*https://oreil.ly/hUpmo*) setting. This ensures the server nodes can communicate with one another using Raft.

The workload nodes will also use a similar configuration, but the Consul service on those nodes will be running in client mode. Consul uses the same binary for servers and clients, so all you need to do is set the `server` setting to `false` for Consul to run as a client.

For more information on running Consul in production, start with the "Create a Local Consul Datacenter" (*https://oreil.ly/EpYZ4*) tutorial.

Next, I'll cover how to interact with your Consul installation.

 The Appendix lists issues you may encounter during the course of completing the exercises throughout the book and how to fix them.

Interacting with Consul

Now that Consul's running, it's time to interact with it. There are three ways to interact with Consul: via its UI, CLI, or API. If running on Kubernetes, you will also use custom resources.

Consul's UI

First, let's look at the UI. In your example installation, the UI is hosted by the Consul servers.

Clients Hosting the UI

It's possible for Consul clients to host the UI, but on Kubernetes, only the servers are configured to host the UI. This is because server pods are more stable—they're less likely to be restarted while you're viewing the UI—whereas client pods come and go as nodes are added and removed from the cluster.

On VMs, Consul operators usually only configure the servers to host the UI for the same reasons.

On VMs, *localhost:8500* is already being forwarded from the VM so you don't need to do anything else to access the UI.

On macOS with minikube, you should already have `minikube tunnel` running to forward ports, but if you're running Kubernetes in the cloud or on Linux, use `kubectl port-forward` to access the UI:

```
$ kubectl port-forward service/consul-ui -n consul 8500
```

To view the UI, navigate to *http://localhost:8500*. You should see the UI, as shown in Figure 3-1.

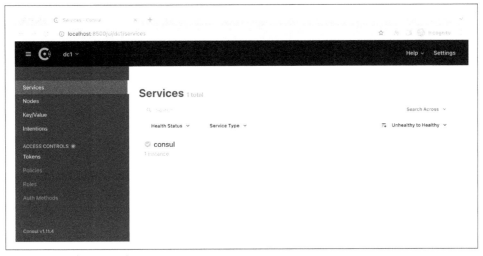

Figure 3-1. The Consul UI

The view that loads by default is the Services view. This view lists all the services Consul has registered. The only service listed right now is the `consul` service. The `consul` service is automatically created by Consul to represent itself. When you register new services in Chapter 4, you'll see more services listed here.

The links on the sidebar are as follows:

Services

Links to the Services view that lists all the services in this datacenter. This is the page you're currently on.

Nodes

Links to the Nodes view that lists all the nodes in this datacenter. On Kubernetes, this will show all your Kubernetes nodes and each Consul server pod. On VMs, this will show all the VMs that Consul clients and servers are deployed on. If you're using Vagrant, then there should only be one node since you're running a single VM.

Key/Value

Links to Consul's key/value store that's used for dynamic service configuration. Consul's key/value store is very powerful, but it's not required when using Consul as a service mesh.[6]

Intentions

Links to Consul's intention management view. Intentions are rules about which services can talk to which other services. Intentions are covered in detail in Chapter 6.

Below these links, there's an Access Controls section with grayed-out links to Tokens, Policies, Roles, and Auth Methods pages. These pages are for configuring Consul's access control list (ACL) system. ACLs set rules about who can perform certain tasks and are a critical part of a secure, production-ready Consul installation. Your current installation does not enable ACLs because ACLs can make it more complicated to try out Consul's features at first. ACLs are covered in Chapter 10.

Across the top, there are the following links:

dc1

This drop-down displays the name of the current Consul datacenter. When deploying multiple Consul datacenters (also described in Chapter 10), the drop-down lets you switch between datacenters.

Help

This drop-down links to Consul's documentation, HashiCorp's Learn site containing tutorials, and Consul's GitHub issue creation page.

Settings

This links to the Settings view used to configure UI settings. These settings only apply to your browser—not to other Consul users.

---

6 To learn more about the key/value store, start with Consul's key/value tutorial (*https://oreil.ly/nHDAO*).

If you navigate back to the Services page and click the `consul` service, you'll be brought into the Service Instances view (Figure 3-2).

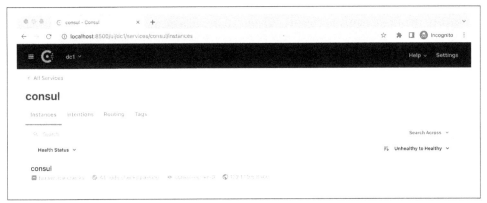

Figure 3-2. The Service Instances view shows all the instances of a specific service—in this case, the `consul` service

This view shows all the running instances of a particular service. For example, if your Kubernetes deployment has multiple replicas, each pod will be listed here. Or, if you are running multiple replicas of the same service across VMs, each replica will be listed here.

In your installation, there is only one instance of the `consul` service listed. This is because Consul registers the `consul` service on each server node, and you're only running one server.

In the Instances list you can see more details about the one instance of the `consul` service:

No service checks
> Service checks are health checks for that service instance. For example, you might register a service check that ensures a certain port is open. The `consul` service has no service checks because it doesn't make sense for Consul to check itself (since if it's down, there is nothing to update the check status). Chapter 8 covers service checks in more detail.

All node checks passing
> Node checks are checks registered to a node instead of a specific service instance. For example, you could register a check on a node that would fail when its disk was full. Node checks affect the health of all service instances on that node. If the node check fails, all services on that node are considered unhealthy.

consul-server-0/vagrant
> This is the name of the node that the instance is running on. By default, the node name is the hostname of the machine where the instance is running. In Kubernetes, the node name is `consul-server-0` because Kubernetes sets the hostname of pods to the pod name. In the Vagrant VM, Vagrant sets the hostname to `vagrant`.

<some-ip-address>:8300
> This is the IP and port of the service instance. The IP will be set to the pod IP in Kubernetes and 127.0.0.1 IP in Vagrant (corresponding to `bind_addr`). The port will be set to 8300 because that's the port for Consul's internal remote procedure call (RPC) traffic.

The Consul UI is a powerful tool for quickly understanding the state of your installation. You'll be using the UI for many tasks later on in the book, but for now let's move on to the CLI.

Consul's CLI

Consul's CLI is invoked by running the consul binary. On Kubernetes, you can use `kubectl exec` to gain a command-line shell inside a Consul pod that has the `consul` binary already installed:

```
$ kubectl exec -it consul-server-0 -n consul -- sh
```

 Type `Ctrl-d` to exit the `kubectl exec` session or run the `exit` command.

On VMs, you're already SSHed into the VM where the `consul` command is available.

View the available subcommands by running `consul --help`:

```
$ consul --help
Usage: consul [--version] [--help] <command> [<args>]
Available commands are:
    acl          Interact with Consul's ACLs
    agent        Runs a Consul agent
    catalog      Interact with the catalog
    config       Interact with Consul's Centralized Configurations
    connect      Interact with Consul Connect
    debug        Records a debugging archive for operators
    event        Fire a new event
...
```

There's not a lot you can do right now with the CLI since your cluster has nothing running, but you can at least list out the current services:

```
$ consul catalog services
```

```
consul
```

And view the current nodes:

```
$ consul catalog nodes
```

```
Node            ID        Address      DC
consul-server-0 4fd4d9c3  172.17.0.9   dc1
minikube        5490f3d5  172.17.0.10  dc1
```

On Vagrant, you'll only see one node listed because you're running a single VM. On Kubernetes, you'll see one node for the Consul server and then one node for each Kubernetes node in your cluster. In minikube, the Consul server is running on the same underlying node as the Consul client, but because it's running in a different pod with a different IP, it's registered as a separate node in Consul.

Under the hood, the CLI is just making HTTP API calls to the local Consul server and displaying the results in a human-readable format. If you want, you can make the same API calls yourself without using the CLI.

Consul's API

You can control almost every resource and action in Consul via its HTTP API. By default, the API is hosted on port 8500. You can make the same API call used by the `consul catalog services` CLI command using `curl` from inside `kubectl exec` or inside the VM:

```
$ curl http://localhost:8500/v1/catalog/services?pretty
```

```
{
    "consul": []
}
```

The path to every API endpoint is prefixed with `/v1/`—for example, `/v1/catalog/services`. Consul uses this prefix to reserve the possibility of adding v2 endpoints later.

Typically, you'll use the UI and CLI to interact with Consul on a day-to-day basis and use the API for more complicated tasks or to get detailed information. The complete reference for Consul's API is available through the Consul API documentation (*https://oreil.ly/Bz12W*).

That ends the tour of Consul's UI, CLI, and API. If you're on Kubernetes, run `exit` to end the `kubectl exec` session:

```
$ exit
```

Summary

This chapter taught you how to deploy Consul on Kubernetes or VMs. To deploy on Kubernetes, you learned how to use the consul-k8s CLI and configure your installation using a *values.yaml* file. On VMs, you learned how to configure Consul using HCL config files and manage the Consul process using `systemd`.

Once Consul was up and running, you used the UI, CLI, and API to interact with Consul and view information about your installation.

Now that Consul is operational, you're ready to learn how to deploy services into the Consul service mesh.

## CHAPTER 4

Adding Services to the Mesh

In the previous chapter, you learned how to deploy Consul on Kubernetes or VMs. The next step to using Consul is to add your services into the mesh. From there, you will be able to use Consul to increase security, observability, and reliability.

A service is "added to the mesh" by registering it with Consul, deploying its sidecar proxy, and ensuring that all communication is routed through that proxy. On Kubernetes, this is done via a simple pod annotation. On VMs, you have a bit more work to do: you need to register your service with Consul, deploy its sidecar proxy, and configure routing.

This chapter and subsequent chapters use an example application I built for the book to illustrate Consul's functionality. After following the exercises with the example application, you will be well equipped to add your own services to the mesh.

Birdwatcher Example Service

To try out the service mesh, you first need some services. You're going to deploy a basic application called Birdwatcher. Birdwatcher consists of two microservices: frontend and backend.

The frontend service hosts a UI as shown in Figure 4-1. Every time you click Shuffle, a new bird is shown.

Figure 4-1. The frontend service hosts the UI for the Birdwatcher service

The frontend service uses the backend service's API to retrieve new birds. Whenever frontend makes a call to backend's /bird endpoint, it gets a JSON response with bird data:

```
{
  "metadata": {
    "hostname": "...",
    "version": "v1"
  },
  "response": {
    "name": "Crimson fruitcrow",
    "imageURL": "...",
    "extract": "The crimson fruitcrow..."
  }
}
```

The application works as follows (see Figure 4-2):

1. When you load the frontend service in your browser at path /, it loads the UI. Then, when you click the Shuffle button, the UI makes a request to the frontend service at /shuffle.

2. The frontend service then makes its own request to backend to retrieve data about a new bird at /bird.

3. The `backend` service responds with a JSON document describing a bird.

4. The `frontend` service forwards the response back to the UI, and the UI displays the new bird.

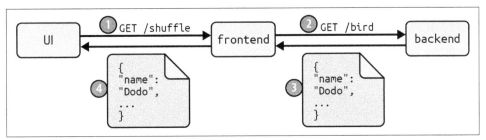

Figure 4-2. Sequence of requests to load a bird in the Birdwatcher application

 `frontend` and `backend` happen to be written in Go, but it doesn't matter to Consul what language your services are written in. As long as they use a network protocol like TCP, HTTP, or gRPC, they can be used in the service mesh.

This example application is meant to illustrate a simple microservices architecture with a UI service and an API service. Now that you understand the architecture of the Birdwatcher application, it's time to deploy it. You'll first deploy Birdwatcher without the service mesh and then learn how to add it to the mesh.

I cover Kubernetes first and then move onto VMs ("Deploying Services on VMs" on page 61).

Deploying Services on Kubernetes

Deploying the `frontend` and `backend` services into Kubernetes requires you to create Deployment and Service resources. As described in Chapter 2, a Deployment is a Kubernetes resource type used to deploy multiple replicas of a service and manage its lifecycle. *Services* in Kubernetes are used to configure routing. Specifically, by creating a Service resource, you create a DNS entry that routes to your deployment. For example, when you create a Service resource for the `backend` service, the `frontend` service can now call `backend` using the URL *http://backend*.[1]

To create resources in Kubernetes, you need to use YAML files. In the same directory where you created your *values.yaml* file for your installation, create a new directory called *manifests* and cd into it:

---

1 If the services were in different namespaces, you'd use *http://backend.<namespace>*.

```
$ mkdir manifests
$ cd manifests
```

Inside the *manifests* directory, create four YAML files, *frontend-deployment.yaml* (Example 4-1), *frontend-service.yaml* (Example 4-2), *backend-deployment.yaml* (Example 4-3), and *backend-service.yaml* (Example 4-4).

Example 4-1. frontend-deployment.yaml

```
apiVersion: apps/v1
kind: Deployment
metadata:
  name: frontend
  labels:
    app: frontend
spec:
  replicas: 1
  selector:
    matchLabels:
      app: frontend
  template:
    metadata:
      labels:
        app: frontend
      annotations:
    spec:
      containers:
        - name: frontend
          image: ghcr.io/consul-up/birdwatcher-frontend:1.0.0
          env:
            - name: BIND_ADDR
              value: "0.0.0.0:6060"
            - name: BACKEND_URL
              value: "http://backend"
          ports:
            - containerPort: 6060
```

Example 4-2. frontend-service.yaml

```
apiVersion: v1
kind: Service
metadata:
  name: frontend
  labels:
    app: frontend
spec:
  type: LoadBalancer
  selector:
    app: frontend
  ports:
    - protocol: TCP
```

```
    port: 6060
    targetPort: 6060
```

 If you're not using minikube on macOS, change the type from Load Balancer to ClusterIP so you don't create a public load balancer.

Example 4-3. backend-deployment.yaml

```
apiVersion: apps/v1
kind: Deployment
metadata:
  name: backend
  labels:
    app: backend
spec:
  replicas: 1
  selector:
    matchLabels:
      app: backend
  template:
    metadata:
      labels:
        app: backend
      annotations:
    spec:
      containers:
        - name: backend
          image: ghcr.io/consul-up/birdwatcher-backend:1.0.0
          env:
            - name: BIND_ADDR
              value: "0.0.0.0:7000"
          ports:
            - containerPort: 7000
```

Example 4-4. backend-service.yaml

```
apiVersion: v1
kind: Service
metadata:
  name: backend
  labels:
    app: backend
spec:
  selector:
    app: backend
  ports:
    - protocol: TCP
```

```
    port: 80
    targetPort: 7000
```

Use kubectl apply to apply these resources to Kubernetes:

```
$ kubectl apply -f ./

deployment.apps/backend created
service/backend created
deployment.apps/frontend created
service/frontend created
```

Use the kubectl get command with the --selector flag to view the frontend service's deployment and service:

```
$ kubectl get deployment,service --selector app=frontend

NAME                         READY   UP-TO-DATE   AVAILABLE   AGE
deployment.apps/frontend     1/1     1            1           1s

NAME                TYPE           CLUSTER-IP     EXTERNAL-IP ...
service/frontend    LoadBalancer   10.98.221.236  127.0.0.1   ...
```

deployment.apps/frontend should show 1/1 READY and 1 AVAILABLE, meaning that there is one pod running and the one container in that pod is ready.

Use the same command with a different selector to view the backend service's deployment and service:

```
$ kubectl get deployment,service --selector app=backend

NAME                        READY   UP-TO-DATE   AVAILABLE   AGE
deployment.apps/backend     1/1     1            1           1s

NAME                TYPE        CLUSTER-IP     EXTERNAL-IP   PORT(S)   AGE
service/backend     ClusterIP   10.96.38.122   <none>        80/TCP    1s
```

Now that your services are deployed, you should be able to access the Birdwatcher UI. First, ensure you're running minikube tunnel so that ports are forwarded properly,[2] and then navigate your browser to *http://localhost:6060*. You should see the Birdwatcher UI as shown previously in Figure 4-1 on page 48.

You've successfully deployed the frontend and backend services into Kubernetes and verified that they are communicating as expected. The current architecture is shown in Figure 4-3. You're accessing *http://localhost:6060*, which is forwarded by minikube tunnel into the frontend pod through its Service resource. The frontend pod is then calling backend via its DNS hostname *http://backend*, which is routed to the backend pod via *its* Service resource.

---

2 If running Kubernetes in the cloud or on Linux, use kubectl port-forward service/frontend 6060.

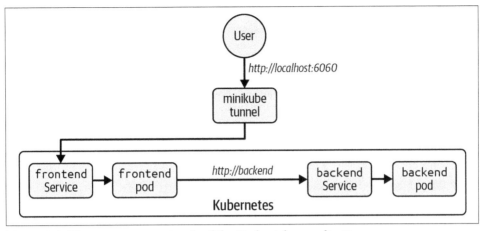

Figure 4-3. The current request path of the Birdwatcher application

You now have a typical set of Kubernetes services running, but these services aren't yet part of the mesh. They're not registered with Consul, and they're not communicating through sidecar proxies. The next step is to add the services to the mesh.

Adding Kubernetes Services to the Mesh

To add a Kubernetes service to the mesh, all you need to do is add the annotation `consul.hashicorp.com/connect-inject: "true"` to the pods. This annotation tells Consul to register the service and inject a sidecar proxy (the sidebar "Under the Hood" on page 55 explains how this works).

You need to edit both deployments to add the annotation. First, edit *frontend-deployment.yaml* and add the annotation under `spec.template.metadata`:

```
# frontend-deployment.yaml
apiVersion: apps/v1
kind: Deployment
# ...
spec:
  # ...
  template:
    metadata:
      labels:
        app: frontend
      annotations:
        consul.hashicorp.com/connect-inject: "true"  ❶
    spec:
      containers:
        # ...
```

❶ Be sure to put quotes around `true`; otherwise you'll get an error.

You must add the annotation to the `spec.template.metadata` key *and not* the top-level `metadata` key. The top-level `metadata` key sets the metadata for the deployment, and the `spec.template.meta data` key sets the metadata for the pods that make up the deployment. Consul is only looking at pods (see "Under the Hood" on page 55), not deployments, and so if the annotation isn't on the pod, Consul won't inject that pod.

Now that you've modified your YAML, you're ready to redeploy the `frontend` service. To do so, use `kubectl apply` in the *manifests* directory:

```
$ kubectl apply -f frontend-deployment.yaml

deployment.apps/frontend configured
```

Kubernetes will handle starting a new pod and spinning down the old pod. You can use the `kubectl rollout` command to wait until the redeployment is complete. It may take a minute or two because Kubernetes needs to download additional Docker images:

```
$ kubectl rollout status --watch deploy/frontend

Waiting for deployment "frontend" rollout to finish...
deployment "frontend" successfully rolled out
```

Now that the pod has been redeployed, when you open up the Consul UI at *http://localhost:8500*, you should see the `frontend` service in the list as shown in Figure 4-4 (ensure you have `minikube tunnel` or `kubectl port-forward` running). This means the `frontend` service is now part of the service mesh!

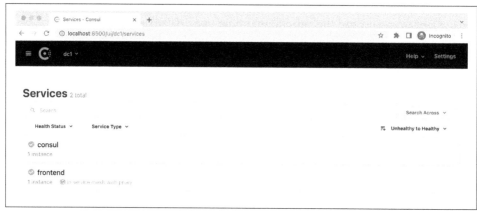

Figure 4-4. The `frontend` service is now listed in the Consul UI

Under the Hood

What happened when you added the `consul.hashicorp.com/connect-inject:` `"true"` annotation?

When you installed Consul, Consul registered a mutating webhook with Kubernetes. A *mutating webhook* is an HTTP callback that allows third-party applications to modify Kubernetes resources.

Whenever a new pod is scheduled, Kubernetes calls Consul's mutating webhook. Consul looks at the pod definition to see if it has the `consul.hashi corp.com/connect-inject:` `"true"` annotation. If so, Consul modifies the pod to add the sidecar proxy (and two init containers).[3]

You can see the modified pod by running `kubectl get pod -l app=frontend -o yaml`. The pod definition will be different from what you have in your local *frontend-deployment.yaml* file because Consul has modified it.

Now add the `backend` service to the mesh by adding the same annotation (and an additional annotation to set some metadata) to *backend-deployment.yaml*:

```
# backend-deployment.yaml
apiVersion: apps/v1
kind: Deployment
# ...
spec:
  # ...
  template:
    metadata:
      labels:
        app: backend
      annotations:
        consul.hashicorp.com/connect-inject: "true"
        consul.hashicorp.com/service-meta-version: "v1" ❶
    spec:
      containers:
        # ...
```

❶ This annotation sets a Consul metadata key. It's not used for now, but in Chapter 9 when you learn about deployment strategies, you'll deploy a `v2` version of the `backend` service.

---

3 An *init container* is a container that runs before the main containers for a pod.

Redeploy it with kubectl apply:

```
$ kubectl apply -f backend-deployment.yaml
```

```
deployment.apps/backend configured
```

Wait for the rollout to complete:

```
$ kubectl rollout status --watch deploy/backend
```

```
Waiting for deployment "backend" rollout to finish:
    1 old replicas are pending termination...
deployment "backend" successfully rolled out
```

Now when you view the UI, you should see both the frontend and backend services listed as in Figure 4-5.

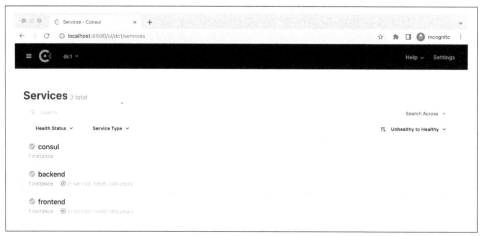

Figure 4-5. Both the frontend *and* backend *services are now also listed in the Consul UI*

Now that both the frontend and backend services are part of the service mesh, if you're using minikube tunnel, try to load the Birdwatcher UI again at *http://local host:6060*. You should see an error similar to Figure 4-6.

You're getting this error because now that the frontend service is part of the mesh, its sidecar proxy is intercepting all incoming traffic. Consul is secure by default and requires all traffic through the service mesh to be authenticated and authorized. When you make a request to the frontend service through your browser, your request is not authenticated or authorized, and so Consul rejects it. This is shown in Figure 4-7.

Figure 4-6. Now when you load the Birdwatcher UI you'll get an error

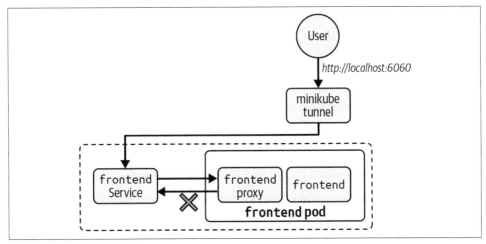

Figure 4-7. The `frontend` pod now has a sidecar proxy that intercepts all traffic. This sidecar rejects incoming traffic that isn't authorized.

If you're using `kubectl port-forward`, you won't see an error. That's because `kubectl port-forward` actually bypasses the sidecar proxy. You can reproduce the same error, however, by using `kubectl exec` to make a normal call to the `frontend` service:

```
$ kubectl exec consul-server-0 -n consul -- \
    curl -sS http://frontend.default:6060

curl: (52) Empty reply from server
command terminated with exit code 52
```

On Windows, if you're using PowerShell, substitute \ for `.

Transparent Proxy

You may be wondering *how* the sidecar proxy is intercepting the traffic. The service appears to still be listening on the same port, so how come traffic is no longer routed to it? Consul calls this feature *transparent proxy*, because the proxy is working without any changes to the service.

When Consul mutated the pod and added the sidecar proxy, it also added two init containers: copy-consul-bin and consul-connect-inject-init.

copy-consul-bin simply copies the Consul binary to a volume so it can be used by consul-connect-inject-init. This is an implementation detail that is required because the consul-connect-inject-init Docker image doesn't have the Consul binary.

consul-connect-inject-init performs several setup tasks, one of which is to write a series of iptables rules. *iptables* is a Linux networking utility that can control how network packets are routed. Consul's iptables rules instruct Linux to route all packets entering and leaving the pod to the sidecar proxy. This is how the proxy intercepts all the traffic.

In the next chapter, you'll deploy an ingress gateway to allow you to call the frontend service with the proper authorization. For now though, you still want to verify that traffic between the frontend and backend services is working as expected. To do so, you can bypass frontend's proxy by using kubectl exec to run a command from inside the frontend container and call backend directly:

```
$ kubectl exec deploy/frontend -c frontend -- \
    curl -si http://backend/bird


HTTP/1.1 200 OK
...
```

If everything works as expected, you should receive an HTTP 200 OK response from the backend service. This request is now going through the service mesh as shown in Figure 4-8.

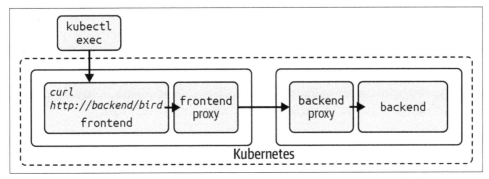

Figure 4-8. kubectl exec bypasses frontend's proxy because it runs inside the front end container. The request to backend then passes through the service mesh as expected.

No-Downtime Mesh Migration

Since Consul is secure by default, all services in the mesh can only send and receive traffic from other mesh services. This can make it difficult to migrate one service at a time to the mesh without downtime because once a service is in the mesh, it can't reach non-mesh services, and vice versa. For example, in the exercise you just completed, there was a brief period of time once frontend was added to the mesh where it could no longer communicate with backend because backend wasn't yet in the mesh.

To perform a no-downtime migration, you can create a Mesh custom resource[4] and set meshDestinationsOnly to false:

```
apiVersion: consul.hashicorp.com/v1alpha1
kind: Mesh
metadata:
  name: mesh
spec:
  transparentProxy:
    meshDestinationsOnly: false
```

This will allow mesh services to call non-mesh services.

In addition, to allow non-mesh services to call mesh services, Consul supports the annotation consul.hashicorp.com/transparent-proxy-exclude-inbound-ports, which can be set to the list of ports your service listens on. This will exclude those ports from transparent proxying, which means non-mesh services can continue to use those ports.

---

4 Custom resources are covered in Chapter 5.

Once all your services have been migrated to the mesh, you can turn `mesh DestinationsOnly` to `true` and remove the port annotations. This will enforce that all future communication must run through the mesh.[5]

To prove that requests are going through the service mesh, you can use the Consul UI's topology view that shows metrics emitted from the sidecar proxies.

Run the previous `kubectl exec` command a couple more times to generate more metrics and then navigate to the topology page for the `backend` service by clicking the `backend` service from the All Services view or navigating to *http:// localhost:8500/ui/dc1/services/backend/topology*. You should see something similar to Figure 4-9 (it may take a minute for the metrics to show up).

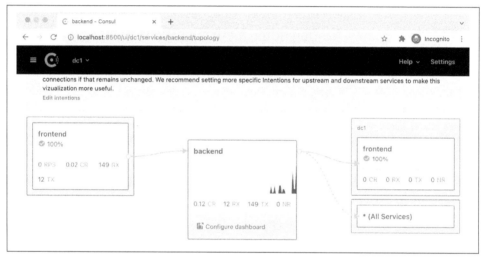

Figure 4-9. The `backend` service's topology view should show metrics from the incoming requests

The warnings you see in the topology view are because ACLs are not enabled, and you haven't created any intentions. ACLs are covered in Chapter 10 and you'll create intentions in Chapter 6. You'll also note that the topology shows `backend` calling `frontend`, which is incorrect. This is because Consul is showing all allowed connections between services. When you create intentions later to only allow certain connections, the topology view will change.

---

5 If you still have services that can't be in the mesh, you can use a terminating gateway. Terminating gateways are covered in Chapter 10.

If you see metrics in the topology view, you've successfully added your services to the mesh! The metrics prove that the sidecar proxies are capturing traffic.

 If you click on the frontend service, you won't see any metrics. This is because the graph only shows incoming requests, and currently you're not making any requests to the frontend service. In the next chapter, you'll deploy an ingress gateway that allows you to make requests to the frontend service, and then you'll also see its metrics.

In the next section, you'll learn how to add services running on VMs into the mesh. If you're only interested in Kubernetes, skip ahead to the chapter summary.

Deploying Services on VMs

In this section, I cover how to deploy and run the Birdwatcher application on VMs without the mesh. Once Birdwatcher is running, you'll learn how to add it to the mesh. The process for adding Birdwatcher to the mesh will be similar to what you'll follow for your own services.

You're going to run both frontend and backend on the same VM for simplicity. The first step is to get the frontend and backend binaries into the VM. If you're using the prepackaged VM via Vagrant then the binaries are already there. If you're using your own VM then the binaries can be found in the book's GitHub repo (*https://oreil.ly/0BOHh*).

With the binaries in place, you need to start them using systemd, just like you did with Consul. First, create two systemd unit files, *frontend.service* and *backend.service*, in */etc/systemd/system/* (ensure you've run vagrant ssh first).

```
$ sudo touch /etc/systemd/system/frontend.service
$ sudo touch /etc/systemd/system/backend.service
```

Edit */etc/systemd/system/frontend.service* and add the contents from Example 4-5.

 You will need to use sudo to edit files in */etc/systemd/system*—for example, sudo vim /etc/systemd/system/frontend.service.

Example 4-5. /etc/systemd/system/frontend.service

```
[Unit]
Description="Frontend service"

# The service requires the VM's network
# to be configured, e.g., an IP address has been assigned.
Requires=network-online.target
After=network-online.target

[Service]
# ExecStart is the command to run.
ExecStart=/usr/local/bin/frontend

# Restart configures the restart policy. In this case, we
# want to restart the service if it fails.
Restart=on-failure

# Environment sets environment variables.
# We will set the frontend service to listen
# on port 6060.
Environment=BIND_ADDR=0.0.0.0:6060

# We set BACKEND_URL to http://localhost:7000 because
# that's the port we'll run our backend service on.
Environment=BACKEND_URL=http://localhost:7000

# The Install section configures this service to start
# automatically if the VM reboots.
[Install]
WantedBy=multi-user.target
```

Next, edit */etc/systemd/system/backend.service* to match Example 4-6.

Example 4-6. /etc/systemd/system/backend.service

```
[Unit]
Description="Backend service"
Requires=network-online.target
After=network-online.target

[Service]
ExecStart=/usr/local/bin/backend
Restart=on-failure

# We will set the backend service to listen
# on port 7000.
Environment=BIND_ADDR=0.0.0.0:7000

[Install]
WantedBy=multi-user.target
```

With the unit files in place, you're ready to enable[6] and start your services:

```
$ sudo systemctl enable frontend backend

Created symlink ...
Created symlink ...
```

Now start the services:

```
$ sudo systemctl start frontend backend
```

 You're running all these steps manually for this exercise, but for a production deployment you should use a provisioning tool like Ansible or Terraform and Packer.

You can check their statuses with `systemctl status`:

```
$ sudo systemctl status frontend backend

  frontend.service - "Frontend service"
     Loaded: loaded...
     Active: active (running)...
...
  backend.service - "Backend service"
     Loaded: loaded...
     Active: active (running)...
```

Now that `frontend` and `backend` are running, you should be able to access the Bird-watcher application by navigating your browser to *http://localhost:6060*. You should see the UI as previously shown in Figure 4-1 on page 48.

The current architecture is shown in Figure 4-10.

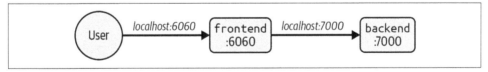

Figure 4-10. frontend listens on port 6060, and backend listens on port 7000

Your services are now running, but they're not yet registered in the mesh and their traffic isn't flowing through sidecar proxies. In the next sections, you'll register the services with Consul and start the sidecar proxies. These are the same steps you'll need to follow for your own services.

---

6 `systemctl enable` sets the services to restart when the VM reboots.

Registering VM Services with Consul

You must register services with Consul so it knows what's running and on what IPs and ports. Services can be registered via configuration files or using Consul's API. In dynamic systems like container orchestrators, it makes sense to use the API, but on VMs it's often simpler to use configuration files because you usually know at provisioning time which services will be running on a particular VM.

Create the service configuration files in Consul's configuration directory */etc/consul.d:*[7]

```
$ sudo touch /etc/consul.d/frontend.hcl
$ sudo touch /etc/consul.d/backend.hcl
```

Add the contents from Example 4-7 to */etc/consul.d/frontend.hcl*.

Example 4-7. /etc/consul.d/frontend.hcl

```
service {
  name = "frontend"

  # frontend runs on port 6060.
  port = 6060

  # The "connect" stanza configures service mesh
  # features.
  connect {
    sidecar_service {
      # frontend's proxy will listen on port 21000.
      port = 21000

      proxy {
        # The "upstreams" stanza configures
        # which ports the sidecar proxy will expose
        # and what services they'll route to.
        upstreams = [
          {
            # Here you're configuring the sidecar proxy to
            # proxy port 6001 to the backend service.
            destination_name = "backend"
            local_bind_port  = 6001
          }
        ]
      }
    }
  }
}
```

---

7 You can place your service configuration in any directory as long as the directory is passed to Consul using the `-config-dir` flag.

```
  }
}
```

 You usually won't set the proxy's port and instead let Consul dynamically allocate it. In this exercise, you're hardcoding the frontend proxy's port to 21000 so it's clear what proxies are on what ports throughout the book.

For */etc/consul.d/backend.hcl*, use the code shown in Example 4-8.

Example 4-8. /etc/consul.d/backend.hcl

```
service {
  name = "backend"
  # backend runs on port 7000.
  port = 7000

  meta {
    version = "v1"
  }

  # The backend service doesn't call
  # any other services so it doesn't
  # need an "upstreams" stanza.
  #
  # The connect stanza is still required to
  # indicate that it needs a sidecar proxy.
  connect {
    sidecar_service {
      # backend's proxy will listen on port 22000.
      port = 22000
    }
  }
}
```

Your own services will need similar configuration files. You'll need to set the name of the service, the port it listens on, and any upstream dependencies (see "Upstreams" on page 66).

You need to tell Consul to reload its configuration so it picks up the new files. To do so, use the consul reload command:

```
$ consul reload

Configuration reload triggered
```

You can use the `consul catalog services` command to check if your services were registered successfully:

```
$ consul catalog services
```

```
backend
backend-sidecar-proxy
consul
frontend
frontend-sidecar-proxy
```

You should see frontend and backend listed, along with their sidecar proxies.

 Why are `frontend-sidecar-proxy` and `backend-sidecar-proxy` listed? Under the hood, Consul treats the sidecar proxies as separate services. You won't see the proxies listed in the UI because they're hidden, but you will in the API or CLI.

Upstreams

You'll note that the frontend service config file lists a set of upstreams and the backend service does not. An *upstream* is a service that is a dependency. Each service that your service needs to call must be defined as an upstream. Because the frontend service needs to call the backend service, it must list backend as an upstream. The backend service doesn't make requests to any other service, so it has no upstreams.

In addition to the upstream service name, you must specify a `local_bind_port`. This isn't the port that the upstream service is listening on. Instead, this is the port that the local sidecar proxy will be listening on, in order to route requests *to that* upstream. Your service, in this example the frontend service, must then make requests to that specific port in order to talk to the specific upstream service, in this example the backend service. This is likely a bit confusing but it will make more sense after the sidecar proxies are deployed.

In the UI at *http://localhost:8500*, frontend and backend should be listed (Figure 4-11), though they will be shown as unhealthy. This is because their sidecar proxies are not yet running.

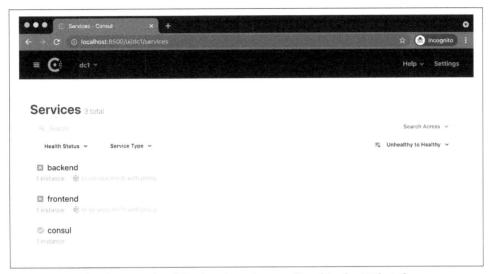

Figure 4-11. The `frontend` and `backend` services are listed in the UI but shown as unhealthy because their sidecar proxies are not yet running

Now that your services are registered in Consul, it's time to deploy their sidecar proxies.

Deploying Sidecar Proxies on VMs

For your services to become healthy, their sidecar proxies must be running. The sidecar proxies run Envoy, which is already installed on the Vagrant VM. If you're running your own VMs, see Envoy's install instructions (*https://oreil.ly/ctvLT*).

Like with the `frontend` and `backend` services, you'll configure a `systemd` service to run each sidecar proxy. The command that `systemd` will use to run the sidecar proxies is:

```
consul connect envoy
```

This command accepts a flag, `-sidecar-for`, that is used to configure the Envoy proxy for a specific service.

 Consul doesn't support running a single Envoy proxy for multiple services because each Envoy proxy must encode the identity of the source service into every request. In addition, you will want to configure different rules for each service's proxy.

First, create the two unit files:

```
$ sudo touch /etc/systemd/system/frontend-sidecar-proxy.service
$ sudo touch /etc/systemd/system/backend-sidecar-proxy.service
```

/etc/systemd/system/frontend-sidecar-proxy.service should match Example 4-9.

Example 4-9. /etc/systemd/system/frontend-sidecar-proxy.service

```
[Unit]
Description="Frontend sidecar proxy service"
Requires=network-online.target
After=network-online.target

[Service]
ExecStart=/usr/bin/consul connect envoy -sidecar-for frontend \
  -admin-bind 127.0.0.1:19000
Restart=on-failure

[Install]
WantedBy=multi-user.target
```

And */etc/systemd/system/backend-sidecar-proxy.service* should match Example 4-10.

Example 4-10. /etc/systemd/system/backend-sidecar-proxy.service

```
[Unit]
Description="Backend sidecar proxy service"
Requires=network-online.target
After=network-online.target

[Service]
ExecStart=/usr/bin/consul connect envoy -sidecar-for backend \
  -admin-bind 127.0.0.1:19001
Restart=on-failure

[Install]
WantedBy=multi-user.target
```

Next, enable the two services:

```
$ sudo systemctl enable frontend-sidecar-proxy
$ sudo systemctl enable backend-sidecar-proxy
```

Now start them up!

```
$ sudo systemctl start frontend-sidecar-proxy
$ sudo systemctl start backend-sidecar-proxy
```

Check that their statuses show active and running:

```
$ sudo systemctl status frontend-sidecar-proxy

frontend-sidecar-proxy.service - "Frontend sidecar proxy service"
    Loaded: loaded...
    Active: active (running)...

$ sudo systemctl status backend-sidecar-proxy

backend-sidecar-proxy.service - "Backend sidecar proxy service"
    Loaded: loaded...
    Active: active (running)...
```

With the proxies now running, the UI should show everything as healthy (see Figure 4-12).

Figure 4-12. The `frontend` and `backend` services are now healthy because their sidecar proxies are running

Reload the Birdwatcher UI at *http://localhost:6060* to see if anything's changed. The UI should look and work the same as before, so how do you know if traffic is being routed through the service mesh?

One way to test if traffic is routing through the mesh is to create an intention that denies all traffic. *Intentions* are authorization rules dictating which services are allowed to communicate. Intentions are covered in detail in Chapter 6, but you can use them now as a quick test to see if traffic is routing through the mesh.

Create an intention to deny all traffic between services using the Consul CLI:

```
$ consul intention create -deny '*' '*'
```

```
Created: * => * (deny)
```

Now try reloading the Birdwatcher UI again. The UI should load a new bird, even though the intention should be denying traffic, so what's going on?

Unlike in Kubernetes, when adding services to the service mesh on VMs you need to make some changes to the URLs your services are using so that they route traffic *through* their sidecar proxies. Right now, frontend is still calling backend directly and bypassing both its local proxy and backend's proxy as shown in Figure 4-13. Intentions are enforced by the proxies, so that's why the Birdwatcher UI is still loading birds.

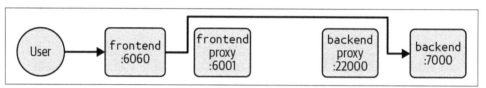

Figure 4-13. frontend is calling backend directly—bypassing the sidecar proxies

Configuring Routing on VMs

You need to configure frontend to route requests to backend through its sidecar proxy. Remember when you configured frontend's upstreams? You set the local_bind_port for the backend service to 6001. Now all you need to do is configure frontend to use that port.

Luckily, the frontend service exposes the environment variable BACKEND_URL for configuring the URL to use to call backend.

 In your own services, you will likely have the same ability to configure upstream URLs via an environment variable or config file. If not, you will have to modify your service code.

Currently in your *frontend.service* file, this environment variable is set to http://localhost:7000, so all you need to do is set it to http://localhost:6001.

Edit */etc/systemd/system/frontend.service* using sudo (*frontend.service*, not *frontend-sidecar-proxy.service*) and change the BACKEND_URL environment variable to http://localhost:6001:

```
...
Environment=BACKEND_URL=http://localhost:6001
...
```

Now run `systemctl daemon-reload` so that `systemd` loads the updated configuration and then use `systemctl restart` to restart the `frontend` service:

```
$ sudo systemctl daemon-reload
$ sudo systemctl restart frontend
```

If you load the Birdwatcher UI at *http://localhost:6060* again, you should now see an error:

```
Unable to call backend: Get "http://localhost:6001/bird": EOF
```

This error confirms that `frontend` is trying to call `backend` through its sidecar proxy and the call is failing because `backend`'s sidecar proxy is rejecting the request.

Delete the intention to allow traffic:

```
$ consul intention delete '*' '*'
```

```
Intention deleted.
```

Go back to the UI and click the Shuffle button. The birds should start loading again!

Now you've confirmed that the requests between `frontend` and `backend` are being routed through the service mesh![8] Figure 4-14 shows what the architecture looks like now.

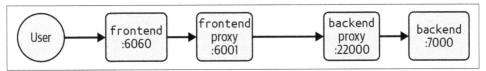

Figure 4-14. frontend is calling backend through the mesh sidecar proxies. You are still accessing frontend directly.

There are two sets of ports for a sidecar proxy: upstream and public ports. A service uses its proxy's upstream ports to route to its dependencies. In this example, 6001 is an upstream port.

Public ports are the ports that incoming traffic from other sidecar proxies will be received on. As shown in Figure 4-14, the public port for `backend`'s sidecar proxy is 22000. Consul configures all proxies so they know the public ports for the other proxies. This is how `frontend`'s proxy can send traffic to `backend`'s proxy.

---

8 Technically, not all requests are traveling through the service mesh. Your browser is still connecting directly with the `frontend` service, not its proxy. In the next chapter you'll add an ingress gateway to fix this.

Summary

In this chapter, you learned how to add services to the Consul service mesh. On Kubernetes, this was a simple matter of annotating your pods with the `consul.hashi corp.com/inject: "true"` annotation. Under the hood, Consul handled automatically injecting the sidecar proxy and intercepting traffic.

On VMs, you registered the services with Consul via configuration files and ran the sidecar proxies using `systemd`. You then learned how to configure services to talk to their upstream dependencies through their local sidecar proxies.

Your services are now communicating through the mesh, but you haven't learned how to expose them to callers that aren't in the mesh. That's why you can't load `frontend`'s UI on Kubernetes and why on VMs, you're bypassing the mesh to load the UI.

In the next chapter, you'll learn how to use ingress gateways to expose user-facing services.

## CHAPTER 5

Ingress Gateways

Consul service mesh is secure by default. This means that Consul requires all requests to be authorized. Chapter 6 covers authorization in detail, but the long and short of it is that user-facing services can't be accessed directly. Instead, they must be accessed through a gateway that sets the necessary authorization.

This is the purpose of a Consul ingress gateway. *Ingress gateways* take unauthorized requests from outside the service mesh and route the requests to services running securely inside the mesh.[1]

In this chapter, you'll learn how ingress gateways work and how to deploy them on Kubernetes or VMs. You'll then continue the exercise from Chapter 4 and expose the Birdwatcher `frontend` service via an ingress gateway.

Why You Need an Ingress Gateway

Most companies require some of their services to be accessed externally by users or API consumers. For example, an ecommerce company may have many internal services that only receive requests from other internal services, but they will also run public-facing services that are accessed by users. Figure 5-1 shows a typical architecture with a load balancer proxying traffic from the public internet directly through to the public-facing services.

---

1 This type of traffic is sometimes called "north-south" traffic.

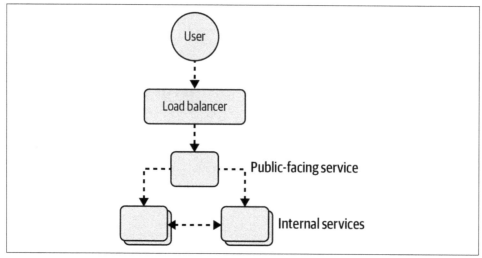

Figure 5-1. A typical architecture with a public-facing service

In this architecture, if the public-facing service is running in the service mesh, it will reject traffic from the load balancer because it is not authorized (see Figure 5-2).[2]

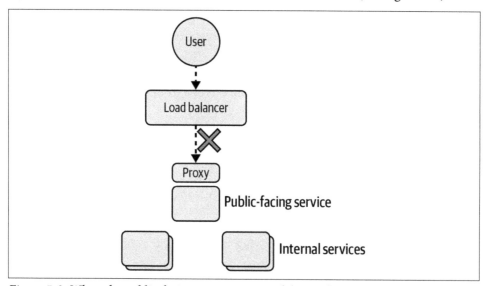

Figure 5-2. When the public-facing service is part of the mesh, it will reject unauthorized traffic

---

2 Requests directly from the load balancer will fail at the first layer of authorization where the client TLS certificate is checked to ensure it is signed by the Consul certificate authority. The full authorization flow is covered in Chapter 6.

This is where the Consul ingress gateway comes into play. When the request is routed through an ingress gateway, it adds in the necessary authorization (see Figure 5-3). In addition, the ingress gateway will emit metrics and allow you to utilize the traffic control and reliability features exposed by Consul.

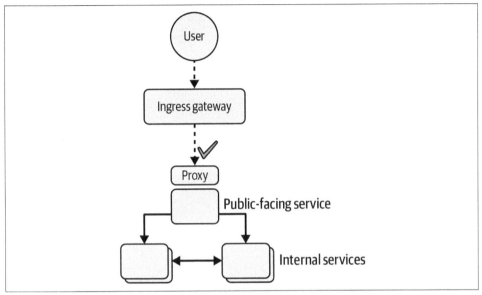

Figure 5-3. If the load balancer is replaced with an ingress gateway, it can authorize requests and forward them to mesh services successfully

Comparison to Kubernetes Ingress Controllers

An ingress gateway is very similar to a Kubernetes ingress controller: they both provide a single entry point to a cluster and can route to multiple services. The key differences are that a Consul ingress gateway ensures traffic is encrypted before forwarding it and that you can use all of Consul's metrics, traffic control, and reliability features.

Consul does support running a sidecar proxy *inside* an ingress controller. This adds an additional process to the controller but allows you to continue using your existing ingress controller while also providing all of Consul's service mesh features. See Consul's ingress controller documentation (*https://oreil.ly/N3cRW*) for more details.

 Under the hood, ingress gateways are Envoy proxies, just like sidecar proxies, but configured differently.

Now that you understand why an ingress gateway is necessary, you're ready to learn how to deploy one. I explain how to deploy an ingress gateway first on Kubernetes and then on VMs ("Deploying an Ingress Gateway on VMs" on page 78).

Deploying an Ingress Gateway on Kubernetes

To deploy an ingress gateway on Kubernetes, you must update your *values.yaml* file and run `consul-k8s upgrade`.

Add the following to the bottom of your *values.yaml* file:

```
ingressGateways:
  # Enable ingress gateways.
  enabled: true
  defaults:
    # Since there is only one node in minikube,
    # set affinity to null so during an upgrade
    # two ingress gateway pods can run on the
    # same node.
    affinity: null

  gateways:
    - name: ingress-gateway
      service:
        # Set the service type to LoadBalancer so
        # that you can access it through minikube tunnel.
        type: LoadBalancer
        # Access the gateway on port 8080.
        ports:
          - port: 8080
      # In production scenarios you should run multiple
      # replicas for redundancy.
      replicas: 1
```

 If running a Kubernetes cluster in the cloud or on Linux with minikube, change the service type of the ingress gateway to `ClusterIP` instead of `LoadBalancer` in *values.yaml*.

In the cloud, leaving it as `LoadBalancer` might expose your cluster to the public internet, which is insecure. On Linux, there's no reason to set it to `LoadBalancer` because you won't be using `minikube` tunnel.

After adding the ingress gateways section, your full *values.yaml* file should look like Example 5-1.

Example 5-1. values.yaml (with comments removed)

```yaml
global:
  name: consul
  metrics:
    enabled: true
  image: hashicorp/consul:1.11.5
  imageEnvoy: envoyproxy/envoy:v1.20.2

server:
  replicas: 1

connectInject:
  enabled: true

controller:
  enabled: true

prometheus:
  enabled: true

ui:
  service:
    type: LoadBalancer
    port:
      http: 8500

ingressGateways:
  enabled: true
  defaults:
    affinity: null
  gateways:
    - name: ingress-gateway
      service:
        type: LoadBalancer
        ports:
          - port: 8080
      replicas: 1
```

Before you deploy the ingress gateway, on macOS start the minikube tunnel command in a new terminal window (if you don't have it running already):

```
$ minikube tunnel
```

In minikube, the tunnel is required for the ingress gateway to start successfully. Without it, Kubernetes will not assign the ingress gateway service an external IP address.[3]

With `minikube tunnel` running, you're ready to deploy the ingress gateway by running `consul-k8s upgrade`:

```
$ consul-k8s upgrade -config-file values.yaml
```

It should take a minute or two for the upgrade to complete.

After the upgrade, the Consul UI at *http://localhost:8500* will show the ingress gateway (see Figure 5-4), but it will be listed as having 0 upstreams. This is because you haven't yet configured it to route to any services.

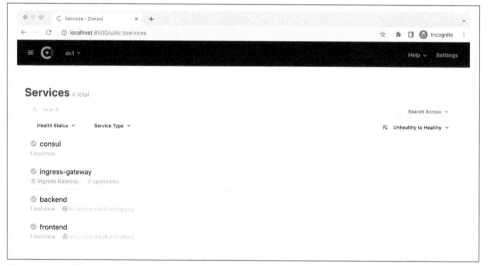

Figure 5-4. The ingress gateway is now listed in the Consul UI

Skip ahead to "Config Entries" on page 82 to learn how to configure your ingress gateway to route to services.

Deploying an Ingress Gateway on VMs

Deploying an ingress gateway on VMs is a two-step process:

---

3 `minikube tunnel` is acting like a cloud provider, but instead of spinning up an actual load balancer it sets the external IP to 127.0.0.1.

1. Register the ingress gateway with Consul.

2. Create a `systemd` service to run the gateway.

The ingress gateway must be registered with Consul first so that when it starts, it can read its configuration from Consul. If you try to start the gateway before it's registered with Consul, it will error. You register an ingress gateway the same way as regular Consul services: by creating a service configuration file inside */etc/consul.d*.

Create */etc/consul.d/ingress-gateway.hcl*:

```
$ sudo touch /etc/consul.d/ingress-gateway.hcl
```

Add the contents from Example 5-2.

Example 5-2. /etc/consul.d/ingress-gateway.hcl

```
service {
  name   = "ingress-gateway"
  kind   = "ingress-gateway"
  port   = 20000 ❶
  checks = [ ❷
    {
      name    = "ingress-gateway listening"
      tcp     = "localhost:20000"
      interval = "10s"
    }
  ]
}
```

❶ Ingress gateways can listen on multiple ports, but they must have one port specified in their service registration for their health check.

❷ Configures a health check to ensure the gateway is up. Consul will check that it can open up a TCP connection to port 20000 every 10 seconds. Health checks are covered in detail in Chapter 8.

Now run `consul reload` so Consul picks up the new file:

```
$ consul reload

Configuration reload triggered
```

In a few seconds, Consul's UI at *http://localhost:8500* should list the ingress gateway (see Figure 5-5). It will show as unhealthy because it's not running yet, so nothing is listening on port 20000.

Figure 5-5. The ingress gateway should be listed in the UI with its status as unhealthy

Run the ingress gateway as another `systemd` service. Create its unit file:

```
$ sudo touch /etc/systemd/system/ingress-gateway.service
```

Edit */etc/systemd/system/ingress-gateway.service* to match Example 5-3.

Example 5-3. /etc/systemd/system/ingress-gateway.service

```
[Unit]
Description="Consul ingress gateway"
Requires=network-online.target
After=network-online.target

[Service]
ExecStart=/usr/bin/consul connect envoy \
    -gateway=ingress \
    -service ingress-gateway \
    -admin-bind 127.0.0.1:19002 \
    -address 127.0.0.1:20000
Restart=on-failure

[Install]
WantedBy=multi-user.target
```

A Deeper Look at the Gateway Command

The command to start the gateway works as follows:

```
/usr/bin/consul connect envoy \ ❶
    -gateway ingress \ ❷
    -service ingress-gateway \ ❸
    -admin-bind 127.0.0.1:19002 \ ❹
    -address 127.0.0.1:20000 ❺
```

❶ The consul connect envoy command is the same as for the sidecar proxies.

❷ The -gateway flag identifies this Envoy process as a gateway of type ingress. This flag supports ingress, mesh, and terminating. Mesh and terminating gateways are covered in Chapter 10.

❸ The -service flag references the ID of the service you previously registered in Consul. When the gateway starts, it will read its configuration from that service registration.

❹ All Envoy proxies host an admin portal that must be given an address to bind to.

❺ The -address flag specifies the address the gateway will be listening on for its health check.

Enable the service:

```
$ sudo systemctl enable ingress-gateway

Created symlink ...
```

Now start the ingress gateway:

```
$ sudo systemctl start ingress-gateway
```

Check that its status is "active (running)":

```
$ sudo systemctl status ingress-gateway

ingress-gateway.service - "Consul ingress gateway"
   Loaded: loaded (...)
   Active: active (running)...
```

Now when you view the Consul UI, the gateway should become healthy after about 10 seconds (see Figure 5-6) because it's listening on port 20000.

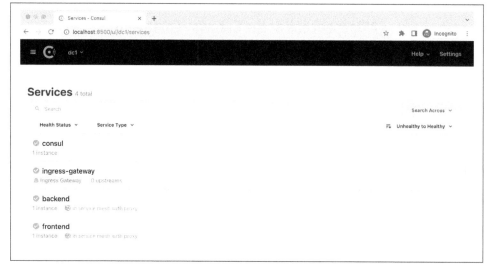

Figure 5-6. The ingress gateway should be listed in the UI with its status as healthy

The ingress gateway will say that it has "0 upstreams" because it's not yet configured to route to any services. Once the gateway is running, you can configure which services it routes to using Consul config entries.

Config Entries

Config entries are sets of Consul configuration that allow you to configure the service mesh dynamically. Config entries are different from the *values.yaml* config file used for Kubernetes and the config files used on VMs because they take effect immediately, without running `consul-k8s upgrade` or reloading Consul.

Config entries are organized into kinds. Each kind has different fields and configures different aspects of the service mesh. These are the kinds you will use in this book:[4]

Ingress gateway
> Configures ingress gateways (as covered in this chapter).

Proxy defaults
> Configures defaults for all proxies in the mesh. Proxy defaults are also covered in this chapter.

Service intentions
> Configures intentions—Consul's authorization rules. Intentions are covered in Chapter 6.

---

4 The complete list of config entries can be found in Consul's documentation (*https://oreil.ly/vgprg*).

Service router, service resolver, service splitter

Part of Consul's traffic routing configuration that will be covered in Chapters 8 and 9.

Config entries are managed differently depending on whether you're running on Kubernetes or VMs, so read on to "Config Entries on Kubernetes", or "Config Entries on VMs" on page 84, respectively.

Config Entries on Kubernetes

On Kubernetes, you manage config entries via custom resources. A *custom resource* is a type of Kubernetes resource that isn't available in a default Kubernetes installation. Kubernetes-native tools like Consul use custom resources to allow users to configure the tool using Kubernetes YAML.

An example custom resource is shown in Example 5-4.

Example 5-4. An example custom resource

```
# ingress-gateway.yaml
apiVersion: consul.hashicorp.com/v1alpha1 ❶
kind: IngressGateway ❷
metadata:
  name: my-gateway
spec: ❸
  listeners: []
```

❶ The `apiVersion` is the version of the schema of this resource. All Kubernetes resources must have an API version.

❷ `kind` specifies the kind of this config entry in camel case.

❸ Each config entry has its own schema and keys that you can set under `spec`. You'll learn about the schemas as you continue through the book.

You can manipulate custom resources using `kubectl` commands just like other Kubernetes resources.

You can create custom resources using `kubectl apply`:

```
$ kubectl apply -f ingress-gateway.yaml

ingressgateway.consul.hashicorp.com/my-gateway created
```

You can list them with kubectl get <kind>:

```
$ kubectl get ingressgateway

NAME        SYNCED   LAST SYNCED   AGE
my-gateway  True     17s           17s
```

 Note the SYNCED column. This column shows whether the config entry is successfully synced to Consul. If the config entry is not synced, you can use kubectl describe to check if there's an error message.

You can edit them using kubectl edit:

```
$ kubectl edit ingressgateway my-gateway
```

ingressgateway.consul.hashicorp.com/my-gateway edited

And you can delete them using kubectl delete:

```
$ kubectl delete ingressgateway my-gateway
```

ingressgateway.consul.hashicorp.com/my-gateway deleted

Now that you understand how config entries work, skip ahead to "Configuring Ingress Gateways" on page 86 to learn how to use config entries to configure your ingress gateway.

Under the Hood

Consul is not aware of Kubernetes custom resources, so another process must translate the Kubernetes resource into a Consul config entry. When you installed Consul, a Kubernetes controller that manages config entries was also installed. A *controller* is a process that communicates with the Kubernetes API and reacts to changes in the cluster.

The config entry controller uses the Kubernetes API to watch for updates to config entry resources. It then syncs those updates to Consul via Consul's HTTP API.

That's why config entry resources have a synced status. If the controller can't sync the resource to Consul, it will set the synced status to false with an error describing why it couldn't sync to Consul.

Config Entries on VMs

On VMs, you can manage config entries via HCL files. An example config entry is shown in Example 5-5.

Example 5-5. An example config entry

```
# ingress-gateway.hcl
Kind = "ingress-gateway"  ❶
Name = "my-gateway"  ❷
Listeners = []  ❸
```

❶ Kind specifies the type of this config entry.

❷ The name of the config entry must be set to the name of the service that is being configured. In this case, an ingress gateway with name my-gateway is being configured.

❸ Kind and Name are fields that exist for all config entries. After that, the supported fields vary depending on the config entry type. In the case of this ingress gateway entry, the field Listeners is supported for setting the ingress gateway listeners.

You can create a config entry using consul config write:

```
$ consul config write ingress-gateway.hcl
```

```
Config entry written: ingress-gateway/my-gateway
```

You can list config entries by their kind using consul config list:

```
$ consul config list -kind ingress-gateway
```

```
my-gateway
```

To update a config entry, you can edit the file and then write it again:

```
$ consul config write ingress-gateway.hcl
```

```
Config entry written: ingress-gateway/my-gateway
```

To delete a config entry, use consul config delete with the flags -kind and -name to refer to a specific entry:

```
$ consul config delete -kind ingress-gateway -name my-gateway
```

```
Config entry deleted: ingress-gateway/my-gateway
```

 Config entries can also be managed via Terraform using the Consul provider (*https://oreil.ly/8812a*).

Configuring Ingress Gateways

Now you understand how to create and edit config entries on Kubernetes or VMs. Before you configure your ingress gateway for a specific use case, it's important to understand how ingress gateway configuration works in general.

Ingress gateways support multiple listeners. A *listener* is a port on which the gateway listens for traffic. Each listener supports different settings:

Port
Which port it's listening on.

Protocol
Which protocol is expected at that port: TCP, HTTP, HTTP2, or gRPC.

TLS
If TLS (think `https://`) is enabled.

Routing configuration
How the listener decides which service to route to. For example, one possible configuration is for the listener to watch for requests with the host header *frontend.example.com* (see the sidebar "The Host Header") and then route those requests to the `frontend` service.

The Host Header

The host header is an HTTP header set to the domain name of the URL that is being called.[5] For example, when you navigate to *www.google.com/search*, your browser sets the header:

```
Host: www.google.com
```

The host header is useful when multiple domain names route to the same gateway in order to find out the domain name used in the request. Ingress gateways can inspect the host header and use it to determine where to route the request.

For example, say you have two services you want to expose, `website` and `billing`, and your users access them at *http://website.example.com* and *http://billing.example.com* respectively. You can configure DNS to route both these domains to your ingress gateway. Then the gateway can inspect the host header and route to either the `website` or `billing` service.

---

5 If a port is specified, then this will also be in the header.

Now that you understand the possible configuration options for an ingress gateway, you can determine the configuration for your specific use case.

In this exercise, you will configure the ingress gateway to route to the `frontend` service as follows:

Port
> The gateway will listen on port 8080. Usually, HTTP services listen on port 80, but since binding to port 80 requires extra privileges, 8080 is often used instead for testing purposes.

Protocol
> The gateway will use the HTTP protocol since that's what the `frontend` service uses.

TLS
> TLS will be disabled so you don't need to deal with certificates, since this is just for testing.

Routing configuration
> The listener will be configured to route to the `frontend` service if the host header is *localhost*. This will allow you to access the ingress gateway on *localhost:8080* using `minikube tunnel` on Kubernetes, or forwarded ports on the Vagrant VM.

Figure 5-7 shows the architecture with the ingress gateway.

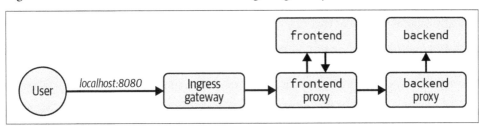

Figure 5-7. Your ingress gateway will listen for traffic on localhost:8080 and route it to the `frontend` *service*

The exact instructions for configuring the ingress gateway depend on whether you're using Kubernetes (refer to the next section) or VMs (see "Configuring Ingress Gateways on VMs" on page 91).

Configuring Ingress Gateways on Kubernetes

To configure your ingress gateway on Kubernetes, you must use an ingress gateway custom resource. Create a file, *ingress-gateway.yaml*, inside the *manifests/* directory with the contents from Example 5-6.

Example 5-6. ingress-gateway.yaml

```
apiVersion: consul.hashicorp.com/v1alpha1
kind: IngressGateway
metadata:
  name: ingress-gateway ❶
  namespace: consul ❷
spec:
  listeners:
    - port: 8080 ❸
      protocol: http ❹
      services: ❺
        - name: frontend ❻
          hosts: ["localhost"] ❼
```

❶ The resource name must match the name of the ingress gateway you set in your *values.yaml* file. In this case, you set it to `ingress-gateway`.

❷ Create it in the consul namespace since that's where the gateway deployment is running.

❸ Your gateway will listen on port 8080.

❹ Your gateway will expect incoming requests to use the HTTP protocol.

❺ The `services` array sets the routing configuration. It configures which services this listener will route to.

❻ Your gateway only needs to route to the `frontend` service.

❼ The `hosts` array configures what host headers will match for this service. Since you will be accessing the ingress gateway through `minikube tunnel` or `kubectl port-forward` on *http://localhost:8080*, this is set to `localhost`.[6]

Apply the custom resource:

```
$ kubectl apply -f ingress-gateway.yaml

ingressgateway.consul.hashicorp.com/ingress-gateway created
```

Ensure that the resource was applied successfully by checking its status with `kubectl get`:

---

6 The host header will actually be set to `localhost:8080`, but the ingress gateway will strip the port before matching.

```
$ kubectl get ingressgateway ingress-gateway -n consul
```

```
NAME               SYNCED   LAST SYNCED   AGE
ingress-gateway    False                  1s
```

 kubectl get ingressgateway ingress-gateway looks weird, but it's the kind of the resource, ingressgateway, followed by the name, which happens to be ingress-gateway in this case.

You should see its SYNCED status set to False. This means that Consul rejected this configuration.

 After applying a Consul config entry, always use kubectl get to ensure it was synced successfully.

To see why it wasn't synced, use kubectl describe:

```
$ kubectl describe ingressgateway ingress-gateway -n consul
...
Status:
  Conditions:
    Last Transition Time:  ...
    Message: writing config entry to consul:
      Unexpected response code: 500 (rpc error making call:
      service "frontend" has protocol "tcp", which does not match
      defined listener protocol "http")
    Reason:                ConsulAgentError
    Status:                False
    Type:                  Synced
  Events:                  <none>
```

The error message shows that the ingress gateway config entry is not syncing because the frontend service uses the TCP protocol, but the listener is set to HTTP. Consul thinks frontend is using the TCP protocol because that's the protocol Consul assumes all services use by default.

Since you know that both your services, frontend and backend, use HTTP, you can set the global default protocol to HTTP instead of TCP. To set the global default protocol, use a proxy defaults config entry.

Create a new file *proxy-defaults.yaml* in *manifests/* with the configuration from Example 5-7.

Example 5-7. proxy-defaults.yaml

```
apiVersion: consul.hashicorp.com/v1alpha1
kind: ProxyDefaults ❶
metadata:
  name: global ❷
  namespace: consul ❸
spec:
  config:
    protocol: http ❹
```

❶ The resource kind is called `ProxyDefaults` because it configures the defaults for all proxies.

❷ There can only be one proxy defaults config for the whole Consul installation, and its name must be `global`.

❸ You could put it in any namespace, but since it's configuring the global defaults, it makes sense in the `consul` namespace.

❹ Set the default protocol to HTTP.

> You can use a ServiceDefaults (*https://oreil.ly/ajDId*) config entry to override the default global protocol for specific services.

Apply the proxy defaults:

```
$ kubectl apply -f proxy-defaults.yaml
```

```
proxydefaults.consul.hashicorp.com/global created
```

Check its status:

```
$ kubectl get proxydefaults global -n consul
```

```
NAME      SYNCED   LAST SYNCED   AGE
global    True     9s            9s
```

It should be synced successfully. Now check the status of the ingress gateway resource:

```
$ kubectl get ingressgateway ingress-gateway -n consul
```

```
NAME               SYNCED   LAST SYNCED   AGE
ingress-gateway    True     9s            1m
```

It should also be synced successfully.[7] Skip ahead to "Testing Out Your Ingress Gateway" on page 93 to see if your config worked!

Configuring Ingress Gateways on VMs

To configure your ingress gateway on VMs, you must use an ingress gateway config entry. First, ensure you're SSHed into the VM:

```
$ vagrant ssh
```

Navigate to the VM's home directory:

```
$ cd ~
```

Next, create a file called *ingress-gateway.hcl*:

```
$ touch ingress-gateway.hcl
```

> This *ingress-gateway.hcl* file is different from the one you used to register the ingress gateway. This file should be created in the home directory, ~.

Add the contents from Example 5-8 to the file.

Example 5-8. ingress-gateway.hcl

```
Kind = "ingress-gateway" ❶
Name = "ingress-gateway" ❷
Listeners = [ ❸
  {
    Port = 8080 ❹
    Protocol = "http" ❺
    Services = [ ❻
      {
        Name = "frontend" ❼
        Hosts = ["localhost"] ❽
      }
    ]
  }
]
```

❶ Kind is the type of the config entry, in this case `ingress-gateway`.

---

7 The config entry controller is constantly trying to resync failed resources. Once the proxy defaults resource set the protocol to HTTP, the config entry controller was able to resync the ingress gateway resource.

❷ The name of the entry must match the name of the ingress gateway service you registered with Consul in the */etc/consul.d/ingress-gateway.hcl* file.

❸ `Listeners` configures the ports the ingress gateway listens on.

❹ Your gateway will listen on port 8080.

❺ Your gateway will expect incoming requests to use the HTTP protocol.

❻ The `Services` array sets the routing configuration. It configures which services this listener will route to.

❼ Your gateway only needs to route to the `frontend` service.

❽ The `Hosts` array configures what host headers will match for this service. Since you will be accessing the ingress gateway through the VM's forwarded port on *http://localhost:8080*, this is set to `localhost`.[8]

 For this book, you're storing your config entry files on the VM, but in a production scenario, you should store your config entry files in version control, so they are backed up.

Now try and write the entry into Consul. You should get an error:

```
$ consul config write ingress-gateway.hcl

Error writing config entry ingress-gateway/ingress-gateway:
  Unexpected response code: 500 (service "frontend" has protocol
  "tcp", which does not match defined listener protocol "http")
```

You're getting an error because the `frontend` service uses the TCP protocol, but the listener is set to HTTP. Consul thinks `frontend` is using the TCP protocol because that's the protocol Consul assumes all services use by default.

Since you know that both your services, `frontend` and `backend`, use HTTP, you can set the global default protocol to HTTP instead of TCP. To do so, you need another config entry kind: proxy defaults.

Create a new file *proxy-defaults.hcl* in the same directory with the configuration from Example 5-9.

---

8 The host header will actually be set to `localhost:8080`, but the ingress gateway will strip the port before matching.

Example 5-9. proxy-defaults.hcl

```
Kind = "proxy-defaults" ❶
Name = "global" ❷
Config {
  protocol = "http" ❸
}
```

❶ The config entry kind is called `proxy-defaults` because it configures the defaults for all proxies.

❷ There can only be one proxy defaults config for the whole Consul installation, and its name must be `global`.

❸ Set the default protocol to HTTP.

> You can use a service-defaults config (*https://oreil.ly/nGW2g*) entry to override the default protocol for specific services.

Write it to Consul:

```
$ consul config write proxy-defaults.hcl
```

```
Config entry written: proxy-defaults/global
```

Now that you've set the default protocol to HTTP, you can retry writing the ingress gateway configuration. This time it should work:

```
$ consul config write ingress-gateway.hcl
```

```
Config entry written: ingress-gateway/ingress-gateway
```

Success! With the ingress gateway config entry written successfully, your ingress gateway is now configured. In the next section, you'll test it out.

Testing Out Your Ingress Gateway

Now that you've successfully deployed and configured your ingress gateway on Kubernetes or VMs, you're ready to test it.

First, check its upstreams in the Consul UI at *http://localhost:8500/ui/dc1/services/ingress-gateway/upstreams*. Consul should show the `frontend` service as an upstream (see Figure 5-8).

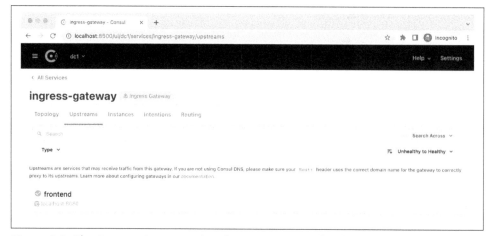

Figure 5-8. The `frontend` service is listed as an upstream for the ingress gateway

This confirms that your ingress gateway has been configured successfully!

You configured the ingress gateway to listen on *localhost:8080* and route requests to the `frontend` service, so to test it, simply navigate your browser to *http://local host:8080*. You should see the Birdwatcher UI!

> If you're using minikube on macOS, ensure `minikube tunnel` is running. If you're running Kubernetes elsewhere, set up a port-forward with `kubectl port-forward service/consul-ingress-gateway -n consul 8080`.

Congratulations! You've successfully deployed and configured your first ingress gateway! Your architecture now looks like that shown in Figure 5-7 on page 87.

On VMs, you'll still be able to bypass the ingress gateway and `frontend`'s sidecar proxy to access the Birdwatcher UI on *localhost:6060*. To prevent this, change `front end`'s bind address so it's not available outside the VM.

Edit */etc/systemd/system/frontend.service* (using `sudo`) and change `Environment =BIND_ADDR=0.0.0.0:6060` to:

```
Environment=BIND_ADDR=127.0.0.1:6060
```

Save the file, and reload `systemd`:

```
$ sudo systemctl daemon-reload
```

Finally, restart the `frontend` service:

```
$ sudo systemctl restart frontend
```

Now you will only be able to access frontend through the ingress gateway on *localhost:8080*.

Ingress Gateways in Production

In the exercise described in this chapter, you are accessing the ingress gateway without TLS encryption. This is fine for testing purposes, but in production, external traffic to your infrastructure must be encrypted using TLS.

Consul ingress gateways support TLS, but as of this writing, they don't support using custom certificates. Instead, they can only use internal certificates. Companies that need to expose services to the public internet need to use their own custom certificates that are signed by an external authority.

To solve this problem, you will either need to:

- Place a load balancer that supports custom TLS certificates in front of the ingress gateway.
- On Kubernetes, use an ingress controller instead of a Consul ingress gateway. See the "Configuring Ingress Controllers with Consul on Kubernetes" web page (*https://oreil.ly/bQoE5*) for more details.
- Use a Consul API gateway. Consul API gateways are similar to ingress gateways, but you configure them using Kubernetes' new Gateway API (*https://oreil.ly/5M1hH*) (a replacement for the Ingress resource), instead of config entries.

  For more information on Consul API gateways, see "Consul API Gateway" on page 228.

Summary

In this chapter, you learned what an ingress gateway is and why it's needed. You learned how to deploy ingress gateways on Kubernetes and VMs and how to configure them. You then deployed your own ingress gateway and configured it to route to the Birdwatcher frontend service.

You're now at a point where your service mesh is fully operational. All your services are registered with Consul and communicating via sidecar proxies. In addition, you have control over external traffic that is routing through the ingress gateway. Now it's time to learn about all the features a service mesh brings to the table. The next chapter focuses on security.

CHAPTER 6
Security

In today's world, security is paramount. Sophisticated attackers are constantly probing for vulnerabilities that can lead to user data being stolen and systems being disrupted. These attacks can ruin a company's reputation and cost hundreds of thousands of dollars. At the same time, securing against these attacks is being made harder as microservices deployments get larger and more complicated—thereby increasing their attack surface.

There are many aspects to securing your systems, and while a service mesh cannot address all of them, it plays an important role. A service mesh implements security improvements via the sidecar proxies that intercept all traffic in and out of services. A service mesh can provide:

- Encryption of traffic between services
- Enforcement of rules about which services can communicate with one another and what kinds of requests are allowed—for example, which HTTP paths can be accessed
- Some mitigation against denial of service attacks by increasing service reliability (covered in Chapter 8)

However, because it operates at the platform level, a service mesh cannot provide:

- Automatic patching of vulnerable libraries
- Elimination of security bugs in service code
- User authentication and authorization (for example, validating passwords)

- Intrusion detection[1]
- Other service-level security improvements

The security improvements a service mesh provides—encryption of traffic between services and enforcement of rules about which services can communicate with one another—are part of implementing a security model known as a *zero trust network*.

This chapter starts with a description of the zero trust network model and why it is an improvement over the older castle and moat model. The chapter then covers how Consul implements a zero trust network and describes TLS encryption—feel free to skip over the TLS sections if you're not interested in understanding Consul at that level.

The chapter ends with an exercise showing how to configure Consul's authorization rules—or *intentions*—that control which services are allowed to communicate.

Zero Trust Networking

Traditional network security architecture followed the castle and moat model. In the *castle and moat* model, services are deployed into an internal private network (the castle) that is not connected to the public internet. A firewall (the moat) secures access to the internal network.[2] Load balancers are deployed outside the private network and are allowed access through the firewall.[3] Because the firewall is in place, it is assumed that everything running inside the internal network can be trusted. For this reason, there's no need for encryption, authentication, or authorization between internal services.

Figure 6-1 shows how traffic is routed through a castle and moat architecture. In step 1, a user's request is routed to a load balancer. The load balancer forwards the request through the firewall to a service running in the trusted internal network (step 2). That service makes a call to its upstream services (step 3). These calls are not encrypted, and the upstream services don't check for authorization because the calls are coming from inside the trusted internal network.

The problem with the castle and moat model is that attackers gain access to the entire system if they breach any point in the internal network, as shown in Figure 6-2. This means that the system's security is only as good as its weakest link.

---

1 Intrusion detection is monitoring for suspicious activity. Consul can help with intrusion detection because it provides metrics about denied requests. If a service suddenly has many denied requests, potentially an attacker is attempting to access it.

2 A firewall is software or hardware that controls access at the edge of the network—where systems are connected to both private and public networks.

3 Like a drawbridge? Maybe that's taking the analogy too far!

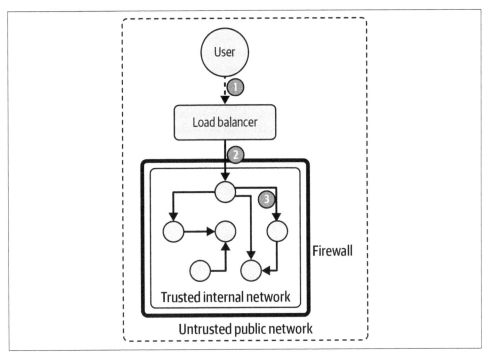

Figure 6-1. Routing inside a castle and moat architecture

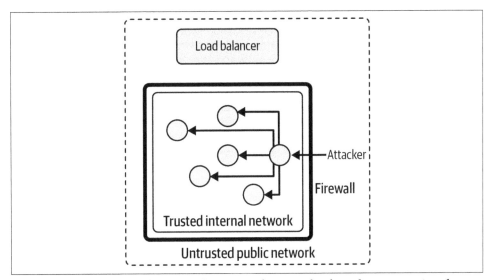

Figure 6-2. In a castle and moat architecture, if an attacker breaches one service, they can make requests to all services and databases inside the network

This is not a theoretical weakness. In 2015, hackers stole millions of sensitive records on government employees from the United States Office of Personnel Management (OPM). The attackers gained access by hacking into the systems of a contracting company whose servers had full access to the OPM network. From there, they were free to exfiltrate the data from the OPM servers.

The weakness of castle and moat systems is unacceptable for most companies, so security engineers recommend a different model: a *zero trust network*.

In a zero trust network, you assume that the internal network is compromised. Therefore, services do not implicitly trust requests simply because they come from inside the internal network. Instead, services implement encryption, authentication, and authorization for all requests.

Encryption
> The process of modifying data such that a third party cannot read the original unmodified data, but the recipient can. In computer networking, the TLS encryption process is often used to ensure that third parties can't intercept and read the data sent between a user and a website they're viewing. TLS is used under the hood when you visit *https://* websites.

Authentication
> The process of validating that an entity is who they claim to be. For example, if a user is claiming to be the admin user, then authentication is the process of checking if the password they provide matches the expected password. Once an entity is authenticated, you still need to check if they're authorized to perform the action they're trying to accomplish.

Authorization
> The process of validating that an authenticated entity is allowed to perform a specific action. For example, a logged-in user may be *authenticated*, but perhaps they're not *authorized* to view the admin panel.

With that context, let's look at how the first requirement in a zero trust network, encryption, is implemented using Consul.

User Versus Service Authorization and Authentication

Service auth (authentication and authorization) is separate from user auth. *Service auth* is about whether a particular service is allowed to make calls to another service. For example, the frontend service is allowed to call the backend service, but it's not allowed to make calls to the credit card database.

User auth is about whether a user is allowed to access a particular resource or perform a certain action. For example, user *isha* is allowed to add items to her shopping cart but not to other users' carts.

A service mesh can help implement service auth, but user auth still needs to be implemented at the service layer because it relies on a lot of business logic that is too complicated to encode into service mesh rules.

Encryption

Encryption is required in a zero trust network because a compromised service can read traffic transmitted between other services—this is known as *traffic sniffing* or a *man-in-the-middle attack* (see Figure 6-3).[4]

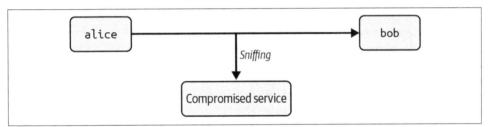

Figure 6-3. A compromised service may be able to intercept traffic sent between two other services

Encryption prevents man-in-the-middle attacks because attackers can't read the data sent between two services. Only the destination service can decrypt the data. As shown in Figure 6-4, if bob is the only service capable of decrypting the traffic from the alice service, then it doesn't matter if attackers intercept the traffic.

---

4 There are many mechanisms for a compromised service to sniff traffic. If the compromised service is on the same node as another service, it may gain permission to inspect the network traffic of other processes. Or a compromised service could use a technique called Address Resolution Protocol (ARP) spoofing to cause another service to mistakenly send its traffic to the compromised service instead of its intended destination.

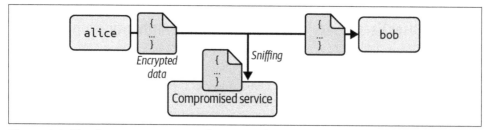

Figure 6-4. The data is now encrypted, so even if the compromised service intercepts the traffic, it cannot obtain the original contents

Consul uses TLS encryption to encrypt traffic between services.

TLS Encryption

TLS encryption has three steps (see Figure 6-5 for an illustration):

1. The source and destination services agree on an encryption key.[5]
2. The source service encrypts its message with the agreed-upon key and sends the encrypted message to the destination service.
3. The destination service decrypts the message using the key agreed upon in step 1.

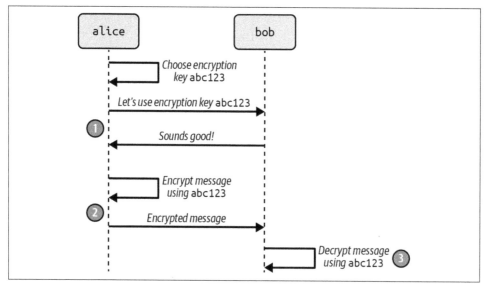

Figure 6-5. An illustration of TLS encryption

---

5 An encryption key is a random sequence of data.

As long as man-in-the-middle attackers don't know the encryption key, they can't decrypt the message. This type of encryption is called *symmetric encryption* because both sides know the encryption key.

But how is the symmetric encryption key agreed upon without an attacker intercepting it? This is where public-key cryptography comes into play.

Public-key cryptography is a method of encryption that utilizes pairs of keys: a private key and a public key. The keys are mathematically linked such that messages encrypted with the public key can be unencrypted only with the private key. Public-key cryptography enables both services to agree upon a symmetric encryption key without that key being sent across the network, where it could be stolen.

Let's walk through an example using two imaginary services, alice and bob, that want to agree upon a symmetric encryption key:

1. The alice service generates two keys, one private, named alice_priv, and one public, named alice_pub.

2. The alice service sends its public key, alice_pub, to the bob service. The key is sent in plain text so it's possible for an attacker to see it.

3. The bob service does the same. It generates two keys, bob_priv and bob_pub, and sends the bob_pub key over to alice. Again, an attacker can see the bob_pub key.

4. Now alice is ready to send the symmetric encryption key to bob. alice encrypts the symmetric encryption key with bob's public key, bob_pub. In algorithmic form this looks like: sym_encryp_key + bob_pub = encrypted_message.

5. alice sends this encrypted message over to bob. Again, it's possible for an attacker to see this encrypted message, but the magic of public-key cryptography is that an attacker cannot decrypt the message using the public keys alice_pub or bob_pub!

   The algorithm used to encrypt the message is designed so that only bob's private key (bob_priv) can decrypt it.

6. bob receives the encrypted message and uses its private key, bob_priv, to decrypt it: encrypted_message + bob_priv = sym_encryp_key.

7. Now that the symmetric encryption key has been securely agreed upon, alice and bob are free to send more messages by encrypting them with the symmetric encryption key.

Figure 6-6 illustrates this process.

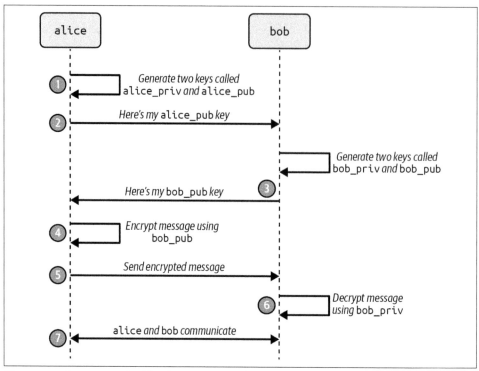

Figure 6-6. Diagram of public-key cryptography

For performance reasons, public-key cryptography is only used at the beginning of a TLS connection to securely agree on a symmetric encryption key. From then onwards, that encryption key is used to encrypt all data sent between the two services.

Now that you understand how TLS encryption works, I will discuss how Consul uses TLS to encrypt service mesh traffic.

Consul Encryption

TLS encryption in Consul is enabled by default. That's what I meant in Chapter 4 when I said that Consul is secure by default.

When a service makes a request through its sidecar proxy, the proxy automatically encrypts it with TLS. The service can continue to make unencrypted requests, and the proxies ensure those requests are encrypted before they leave the local network. This process is shown in Figure 6-7. In step 1, Consul supplies the public and private keys to each proxy. When frontend sends an unencrypted request to backend, the request is intercepted by frontend's proxy (step 2) and encrypts it before forwarding it to backend (step 3). On the incoming side, backend's proxy intercepts the request (step 4) and decrypts it before forwarding it to the destination service (step 5).

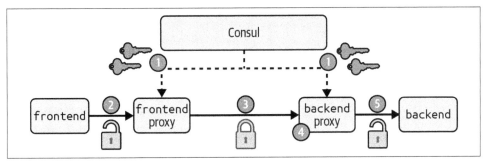

Figure 6-7. Sidecar proxies automatically encrypt traffic with TLS

With this mechanism in place, services don't need to be modified to use TLS—it all happens automatically without their knowledge.

Traffic Between Services and Proxies

You may wonder what's stopping attackers from intercepting traffic on the outgoing request *before* the proxy encrypts it or on the incoming request *after* the proxy decrypts it. On Kubernetes, traffic between the service and its proxy happens over localhost on the pod's internal network since all containers in a pod share the same networking stack. This means the traffic isn't exposed outside of the pod.[6]

On VMs, other services or programs running on the VM could access the localhost network, so it's not as secure. A secure Consul installation on VMs will run only one service per VM or use Linux network namespaces or Unix domain sockets to further isolate the network between a service and its proxy.

Now you understand how Consul implements encryption. Encryption is essential in a zero trust network, but it's not enough. Without authentication and authorization, a compromised service can still make requests through its sidecar proxy to any service in the network.

Authentication

Authentication is the process of verifying that someone is who they claim to be; it's about verifying identity. In service networking, this has two parts: the source service must verify that the destination service is actually the service it means to talk to, and the destination service must verify that the source service is actually who it claims to be.

---

6 However, if a container in the pod was compromised, it could sniff the traffic.

Concretely, if frontend is calling backend, frontend needs to ensure that it's actually talking to backend. If it doesn't check, then it could be sending its requests to an attacker. backend also needs to verify which service is making the request so it can check if that service is allowed to communicate with it.

Consul piggybacks on TLS to also implement authentication. When services exchange their public keys, they actually exchange a certificate. This certificate contains the public key and information about the service, such as its ID.

These are the steps involved in TLS authentication (see Figure 6-8 for an illustration):

1. Consul issues a public certificate and private key to each service. Encoded in the public certificate is the ID of that service.

2. During the key exchange process discussed in "TLS Encryption" on page 102, the public certificates for each service are exchanged.

3. Each service inspects the other service's public certificate and ensures that the service ID is what is expected. For example, the frontend service will verify that the certificate is for the backend service, and the backend service will verify that the certificate is for the frontend service.

4. If the identities are verified, the request will continue.

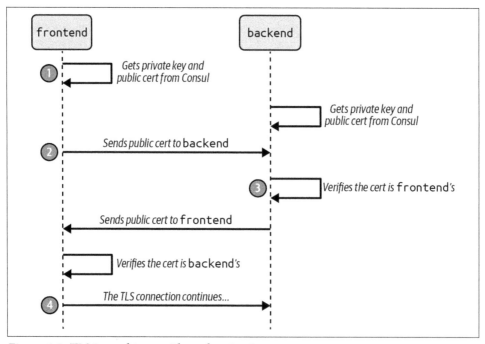

Figure 6-8. TLS is used to provide authentication

This mechanism is known as mutual TLS (mTLS) because the destination service also verifies the certificate of the calling service. In regular TLS—for example, when you navigate to *https://www.google.com*—the destination site doesn't validate your certificate.

SPIFFE

The service ID encoded in the public certificates of each service is a SPIFFE ID. *SPIFFE* (pronounced *spiffy*) stands for Secure Production Identity Framework For Everyone, and it is a set of specifications platforms can follow to issue certificates and attest to the identities of workloads.

The SPIFFE ID is a simple URI. For example, the SPIFFE ID for the `backend` service is:

```
spiffe://<trust-domain>/ns/default/dc/dc1/svc/backend
```

`spiffe://`
Identifies the URI as a SPIFFE ID.

`<trust-domain>`
An ID like `b84acb72-ff87-543f-ef53-51d534f97329.consul`. The first part is a randomly generated universally unique identifier (UUID) that Consul uses to ensure two separate Consul installations don't have the same trust domain. The `.consul` part identifies the trust domain as coming from a Consul cluster. All SPIFFE IDs within a Consul datacenter (and federated datacenters) share the same trust domain.

`ns/default`
Identifies the namespace of the service. In this example, the namespace of the `backend` service is `default`.[7]

`dc/dc1`
Identifies the Consul datacenter. The datacenter `backend` is running in is named `dc1`. Chapter 10 covers Consul datacenters.

`svc/backend`
The name of the service.

When exchanging public certificates, Consul will parse the SPIFFE ID and verify that it matches what it expected.

---

7 Consul namespaces are a Consul Enterprise feature.

You may be wondering what stops an attacker from creating their own public certificate to impersonate any service. To prevent this, Consul acts as a certificate authority. A *certificate authority* is an entity that issues certificates that are cryptographically signed such that they could only have come from that certificate authority. A certificate authority also has a public certificate known as a certificate authority certificate (CA cert). When presented with a service's public certificate—for example, backend's certificate—a third party can use the CA cert to check that the public certificate was signed by that certificate authority.

Concretely, in addition to the public certificate and private key that Consul issues to each service, it also issues each service its CA cert. This CA cert can then be used to verify that Consul actually issued the public certificate and that it's safe to trust the service ID encoded within.

Figure 6-9 shows the entire flow when frontend makes a request to backend.

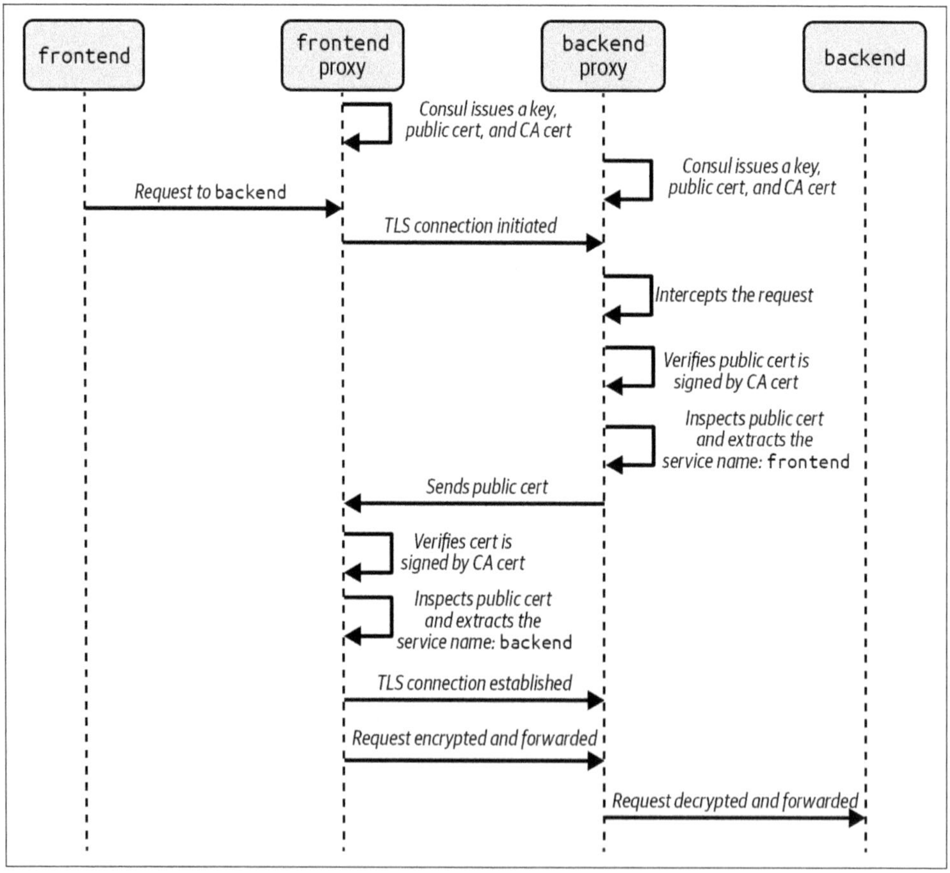

Figure 6-9. Everything that occurs for frontend to talk to backend securely

You can see the public certificates of each service using the `openssl` tool.

On Kubernetes, use `kubectl exec` to connect to backend's sidecar proxy on port 20000 and view its certificate:

```
$ kubectl exec deploy/backend -c backend -- \
    sh -c 'openssl s_client -connect $(hostname -i):20000 | \
    openssl x509 -noout -text'
...
Certificate:
...
    X509v3 Subject Alternative Name: critical
        URI:spiffe://...consul/ns/default/dc/dc1/svc/backend
...
```

On VMs, SSH into the VM:

```
$ vagrant ssh
```

And use the `openssl s_client` command to connect to backend's sidecar proxy on `localhost:22000`:

```
$ openssl s_client -connect localhost:22000 | openssl x509 -noout -text
...
    X509v3 Subject Alternative Name:
        URI:spiffe://....consul/ns/default/dc/dc1/svc/backend
...
```

> The `openssl s_client` command will also output some errors like `unable to get local issuer certificate` and `Can't use SSL_get_servername`. These errors can be safely ignored. They're logged because the Consul certificates aren't public, which is expected because they're only needed for internal mesh communication.

On both Kubernetes and VMs, you should see the SPIFFE ID of the backend service under the `X509v3 Subject Alternative Name` line.

You now understand how Consul leverages TLS to provide authentication and encryption. In the next section, you'll learn how Consul implements the final requirement of a zero trust network: authorization.

Authorization and Intentions

Authorization is the process of determining whether an authenticated entity is allowed to perform a certain action. For example, is this service allowed to make requests to that service, or is it allowed to access a particular HTTP path?

Consul implements authorization via its intentions system. *Intentions* are rules governing which services are allowed to communicate.

Every intention has a source and destination. For example, an intention might allow a specific service `frontend` (source) to connect to a specific service `backend` (destination). Alternatively, a wildcard (*) can be used as the source or destination, allowing, for example, an ingress gateway to connect to any service, or any service to connect to any other service.

Identity-Based Security

A key insight to understand is that Consul provides network security through *identity*. Instead of configuring rules about which *IP addresses* can communicate, like you would with a firewall, you configure rules about which *entities* can communicate.

Identity-based security is a huge improvement over IP-based security because it performs better in dynamic environments where workloads are relocating constantly, it is less complicated to manage, and it works even when requests are traversing network boundaries through proxies and gateways.

In addition to specifying rules based on the source and destination service, you can use intentions to control which HTTP or gRPC methods, paths, and headers are authorized. Consul calls these intentions *application aware* because they concern how the application actually operates. For example, you could create an intention that allows the ingress gateway to access any path on the `frontend` service except for `/admin`.

Application aware intentions require you to configure the destination service's protocol as HTTP or gRPC. You did this already in Chapter 5 via the proxy defaults config entry.

The mechanism for enforcing intentions is as follows:

1. The source service makes a request to the destination service.

2. The destination service's proxy verifies the public certificate of the source service and extracts the service name from the SPIFFE ID.

3. The destination service's proxy checks against its list of intentions to ensure that the source service is authorized to make a connection.

4. If there are application aware intentions set, the destination service's proxy also verifies that that particular request is allowed.

> ## Intention Precedence
>
> If more than one intention applies to a service, Consul will use the intention that has the highest precedence. Intentions that involve all services, represented by *, have lower precedence.
>
> For example, if there are two intentions that apply to the backend service:
>
> 1. frontend => backend (allow)
> 2. * => * (deny)
>
> Then the first intention will take precedence because it doesn't use *.
>
> Consul will also reject intentions at the same level of precedence that conflict with each other. For example, you can't create frontend => backend (deny) if you already have frontend => backend (allow).

You can configure intentions via config entries, or Consul's UI, CLI, or HTTP API.

The most common way to configure intentions is via config entries because this allows the intentions to be written in code and applied as part of a continuous integration and continuous delivery (CI/CD) pipeline. The UI or the CLI is convenient for testing purposes or to try out small changes.

The following sections cover configuring intentions via the UI and config entries since those are the most common methods. If you're interested in using the CLI or API, see Consul's CLI (*https://oreil.ly/3fqQa*) or API (*https://oreil.ly/wXhaX*) docs.

Configuring Intentions with Consul's UI

You can configure intentions through Consul's UI by navigating to the Intentions page at *http://localhost:8500/ui/dc1/intentions* (see Figure 6-10).

Click the Create button to create a new intention that denies traffic from frontend to backend:

1. For Source Service, select frontend.
2. For Destination Service, select backend.
3. For the description, use "Deny frontend to backend."
4. For the action, select Deny.

When you've entered everything correctly (see Figure 6-11), click the Save button.

Figure 6-10. *The Intentions list page*

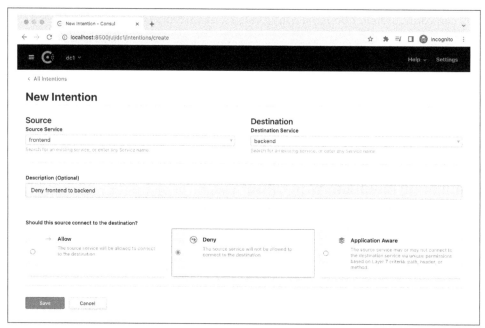

Figure 6-11. *Create an intention denying traffic from* frontend *to* backend

Load the Birdwatcher application at *http://localhost:8080* to test out the intention (if you're running Kubernetes, ensure minikube tunnel or kubectl port-forward is running). You should see an error similar to Figure 6-12 (it may take two reloads). The UI is showing an error because the connection between frontend and backend is no longer allowed.

Figure 6-12. The `frontend` service is unable to call the `backend` service because the connection is denied

On VMs you'll get an error `RBAC: access denied` because the proxies are configured slightly differently on VMs.[8]

RBAC stands for role-based access control. It's the system used internally by Envoy for enforcing intentions.

This proves that the intention is working!

Specifying intentions via the UI is great for quick testing purposes, but it's not suitable for managing intentions long-term:

- The intentions are not codified anywhere, so you can't recover them if they're accidentally deleted.
- There is no version control to see what has changed.

---

8 The nitty gritty is that on Kubernetes, proxies are only sent the addresses for upstream services they're allowed to talk to. This limits how much data is sent to proxies, which improves performance. Since traffic to the `backend` service is not allowed, `frontend`'s proxy isn't sent its address by Consul. When `frontend`'s sidecar proxy intercepts the outgoing request, it sees it's for a service that it doesn't have an address for, so it fails the request immediately, resulting in an end-of-file (EOF) error. On VMs, the request makes it to `backend`'s proxy, which returns the RBAC error response.

- You can't apply them as part of a CI/CD process.

For these reasons, it's best to use config entries to manage intentions for anything other than testing purposes because config entries can be codified and stored in version control systems.

In the following sections, you'll learn how to use config entries to manage intentions, so before you move on, be sure to delete the intention you just created by clicking the three dots to the right of the intention under the Actions column and selecting Delete (see Figure 6-13).

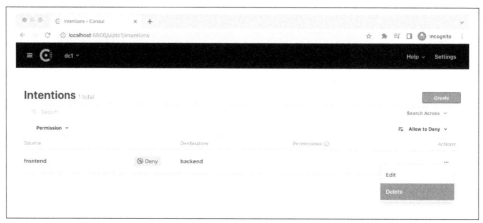

Figure 6-13. Delete the intention through the UI

Configuring Intentions with Config Entries

You learned about config entries in Chapter 5 when you used them to configure the ingress gateway. You can also use config entries to manage intentions via the service intentions config entry.

A service intentions config entry configures multiple intentions for a specific destination. That's why the name of the config entry kind is plural.

First, I cover how to configure intentions on Kubernetes and then on VMs ("Configuring intentions on VMs" on page 118).

Configuring intentions on Kubernetes

Example 6-1 shows a service intentions resource.

Example 6-1. Service intentions resource that denies all traffic

```
# deny-all-service-intentions.yaml
apiVersion: consul.hashicorp.com/v1alpha1
kind: ServiceIntentions
metadata:
  name: deny-all
  namespace: consul
spec:
  destination:
    name: "*" ❶
  sources:
    - name: "*" ❷
      action: deny ❸
```

❶ The name of the destination service. In this example, the destination service is the wildcard * to indicate all services.

❷ Under `spec.sources`, the source services are specified. In this example, the source service is also * to indicate all services.

❸ `action` can be set to either `allow` or `deny` to permit this communication or refuse it. In this example, it's set to `deny`, which means that no services are allowed to communicate.

Create a new file named *deny-all-service-intentions.yaml* in *manifests/* with the contents from Example 6-1 and run:

```
$ kubectl apply -f deny-all-service-intentions.yaml

serviceintentions.consul.hashicorp.com/deny-all created
```

Verify that the intention was synced to Consul:

```
$ kubectl get serviceintentions deny-all -n consul

NAME       SYNCED   LAST SYNCED   AGE
deny-all   True     1s            1s
```

You should be able to view the intention in the Consul UI's intention list as shown in Figure 6-14.

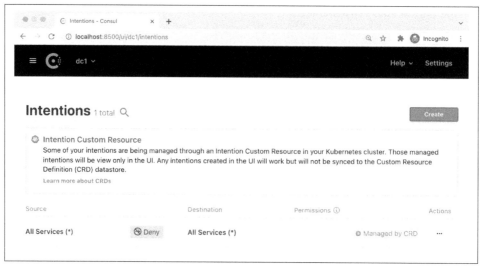

Figure 6-14. The intention you just created is listed on the Intentions page

 The intention has the label "Managed by CRD". The label indicates that the intention was created via a Kubernetes custom resource definition (CRD). Intentions created through CRDs are read-only in the UI because if the UI were to allow the intention to be edited, it would get out of sync with the underlying Kubernetes resource.

Now that all connections are being denied, not only can the frontend service not call the backend service, but the ingress gateway won't be able to call the frontend service as shown in Figure 6-15.

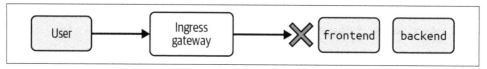

Figure 6-15. The ingress gateway can't talk to the frontend service

To test this, navigate to *http://localhost:8080*. You should see an error:

```
RBAC: access denied
```

 The ingress gateway doesn't reject the incoming request because incoming requests into ingress gateways aren't governed by intentions. Instead, you can use intentions to govern what services the ingress gateway can talk to.

Now, add an intention that allows the ingress gateway to connect with `frontend`. This intention will take precedence over the deny-all intention because it applies to specific services.

Create a new file, *frontend-service-intentions.yaml*, in *manifests/* with contents from Example 6-2.

Example 6-2. frontend-service-intentions.yaml

```yaml
apiVersion: consul.hashicorp.com/v1alpha1
kind: ServiceIntentions
metadata:
  name: frontend
spec:
  destination:
    name: frontend
  sources:
    - name: ingress-gateway
      action: allow
```

Apply the config:

```
$ kubectl apply -f frontend-service-intentions.yaml

serviceintentions.consul.hashicorp.com/frontend created
```

 You're creating this resource in the default namespace because that's where the `frontend` service is deployed. If you wanted to, you could manage all intentions in the consul namespace.

Ensure it is synced:

```
$ kubectl get serviceintentions frontend

NAME       SYNCED   LAST SYNCED   AGE
frontend   True     1s            1s
```

Figure 6-16 shows the current intention rules.

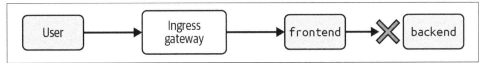

Figure 6-16. The ingress gateway can talk to the `frontend` service, but the `frontend` service is not yet allowed to talk to the `backend` service

Now refresh *http://localhost:8080* twice. The ingress gateway can communicate with the frontend service, so the UI will load, but the backend service should still be unreachable, as shown in Figure 6-12 on page 113.

To allow the frontend service to communicate with backend, create another intention file, *backend-service-intentions.yaml*, in *manifests/* with the contents of Example 6-3.

Example 6-3. backend-service-intentions.yaml

```
apiVersion: consul.hashicorp.com/v1alpha1
kind: ServiceIntentions
metadata:
  name: backend
spec:
  destination:
    name: backend
  sources:
    - name: frontend
      action: allow
```

Apply it to the cluster:

```
$ kubectl apply -f backend-service-intentions.yaml
```

```
serviceintentions.consul.hashicorp.com/backend created
```

Ensure it is synced:

```
$ kubectl get serviceintentions backend
```

```
NAME      SYNCED   LAST SYNCED   AGE
backend   True     1s            1s
```

Refresh your browser a couple of times at *http://localhost:8080* (it may take a few seconds for the connection to be allowed again). The Birdwatcher application should now be fully operational!

Congratulations, you've successfully configured intentions to protect your cluster. Communication between all services is locked down, except for your rules allowing the ingress gateway to talk with frontend, and frontend to talk with backend.

The next section covers configuring intentions on VMs. For Kubernetes users, skip ahead to "Application Aware Intentions" on page 122 to learn about how to further tighten your security by specifying which HTTP paths frontend can access.

Configuring intentions on VMs

Example 6-4 shows an example service intentions resource.

Example 6-4. Service intentions resource that denies all traffic

```
Kind = "service-intentions"
Name = "*" ❶
Sources = [
  {
    Name = "*" ❷
    Action = "deny" ❸
  }
]
```

❶ The name of the destination service. In this example, the destination service is the wildcard * to indicate all services.

❷ Under `Sources`, the source services (those calling the destination service) are specified. In this example, the source service is also * to indicate all services.

❸ `Action` can be either `allow` or `deny` to permit or refuse the communication between the source and destination. In this case, it's set to `deny`, which means that no services are allowed to communicate.

To try out creating a service intentions config entry, first SSH into your VM:

```
$ vagrant ssh
```

Navigate to the home directory inside the VM:

```
$ cd ~
```

Create a new file called *deny-all-service-intentions.hcl* with contents from Example 6-4 and apply the intention with `consul config write`:

```
$ consul config write deny-all-service-intentions.hcl

Config entry written: service-intentions/*
```

Verify the intention was created by reading it back with `consul config read`:

```
$ consul config read -kind service-intentions -name '*'

{
    "Kind": "service-intentions",
    "Name": "*",
    "Sources": [
        {
            "Name": "*",
            "Action": "deny",
            "Precedence": 5,
            "Type": "consul"
        }
    ],
    "CreateIndex": 5031,
```

```
    "ModifyIndex": 5031
}
```

 You can also use the `consul intention list` command to view intentions:

```
$ consul intention list

ID  Source  Action  Destination  Precedence
    *       deny    *            5
```

You should also be able to view the intention in the Consul UI on the Intentions page as shown in Figure 6-17.

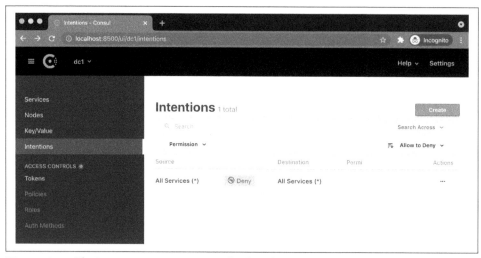

Figure 6-17. The intention you just created is listed on the Intentions page

Now that Consul is denying all connections, not only can `frontend` not call `backend`, but the ingress gateway should not be able to call the `frontend` service. To test this, navigate your browser to *http://localhost:8080* and refresh twice.

You should see the following:

```
RBAC: access denied
```

You need to refresh twice because changes to intentions are enforced on new connections and your first refresh reuses the previous connection.

As expected, the ingress gateway is unable to make a connection with `frontend`.

To enable this connection, add an intention that allows the ingress gateway to connect with frontend. This intention will take precedence over the deny-all intention because it applies to specific services.

Create a file *frontend-service-intentions.hcl* with the contents from Example 6-5.

Example 6-5. frontend-service-intentions.hcl

```
Kind    = "service-intentions"
Name    = "frontend" ❶
Sources = [
  {
    Name   = "ingress-gateway" ❷
    Action = "allow"
  }
]
```

❶ The destination service is frontend.

❷ The source service is the ingress gateway.

Write it to Consul:

```
$ consul config write frontend-service-intentions.hcl

Config entry written: service-intentions/frontend
```

Now refresh your browser twice. The UI should load, which means the request was allowed between the ingress gateway and the frontend service, but you should still get an error because the frontend service is not yet allowed to talk to the backend service (see Figure 6-12 on page 113).

You need to create another intention to allow frontend to talk to backend. Create another file, *backend-service-intentions.hcl*, with the contents from Example 6-6.

Example 6-6. backend-service-intentions.hcl

```
Kind    = "service-intentions"
Name    = "backend"
Sources = [
  {
    Name   = "frontend"
    Action = "allow"
  }
]
```

Write it to Consul:

```
$ consul config write backend-service-intentions.hcl
```

```
Config entry written: service-intentions/backend
```

Now refresh your browser twice, and you should see the birds are back!

Congratulations, you have successfully protected your cluster! Communication between all services is locked down, except for your rules allowing the ingress gateway to talk to `frontend`, and `frontend` to talk to `backend`. In the next section, you'll learn how to lock down permissions even further at the HTTP layer.

Application Aware Intentions

The intentions you created don't specify which HTTP paths or methods each service is allowed to call on the destination service. By default, all paths and methods are allowed if the intention is an allow intention. Imagine a scenario where one of your services has HTTP paths that you don't want accessed from certain services. For example, what if the `frontend` service has an admin panel at `/admin` that you don't want to be exposed through the ingress gateway? Application aware intentions allow you to restrict service access to specific areas of the application.

The Birdwatcher service has just such an admin panel.[9] Navigate to *http://local host:8080/admin*, and you should see the admin panel as shown in Figure 6-18.

Figure 6-18. frontend's admin panel is currently accessible through the ingress gateway

---

9 Well, it doesn't actually administrate anything, but it gives you the idea.

You'd like to lock down access so the ingress gateway can't make requests to /admin or any path that starts with /admin.

To support this use case, you can configure intention rules to match on HTTP request attributes like paths. Specifically, you can match on the following attributes:

- HTTP paths; for example, /admin
- HTTP methods; for example, GET and POST
- HTTP headers; for example, X-Token
- A combination of any of the above

 This also works for gRPC services by matching on the service method.

In order to restrict the ingress gateway's routing to any path other than /admin, you need to modify your frontend intention as follows:

1. All requests matching the path prefix /admin should be denied.
2. All other requests should be allowed.

First, I cover how to do so on Kubernetes and then on VMs ("Configuring application aware intentions on VMs" on page 125).

Configuring application aware intentions on Kubernetes

Currently in *frontend-service-intentions.yaml* under the sources array, you're specifying both the service name and an action:

```
# ...
sources:
- name: ingress-gateway
  action: allow
```

Instead of setting the action key, you can set a permissions key that supports specifying more granular permissions.

Edit *frontend-service-intentions.yaml* and change its contents to match Example 6-7.

Example 6-7. frontend-service-intentions.yaml

```yaml
apiVersion: consul.hashicorp.com/v1alpha1
kind: ServiceIntentions
metadata:
  name: frontend
spec:
  destination:
    name: frontend
  sources:
    - name: ingress-gateway
      permissions: ❶
        - http:
            pathPrefix: /admin ❷
          action: deny
        - http:
            pathPrefix: / ❸
          action: allow
```

❶ Consul supports a `permissions` array in place of the `action` key for specifying more granular permissions.

❷ The first permission matches on all requests with the path prefix `/admin` and sets the action to deny.

❸ The second permission matches all requests using the path prefix `/`. This permission will apply to any request that wasn't matched by the first permission and will allow the request. (If two permissions match the request, the first in the list will apply.)

> The full schema of the service intentions resource can be found in Consul's reference documentation (*https://oreil.ly/TK7Dp*).

Apply the change:

```
$ kubectl apply -f frontend-service-intentions.yaml
```

```
serviceintentions.consul.hashicorp.com/frontend configured
```

Check its synced status:

```
$ kubectl get serviceintentions frontend
```

```
NAME      SYNCED  LAST SYNCED  AGE
frontend  True    1s           10m
```

With the new intention rules in place, you should no longer be able to access the /admin path. Refresh *http://localhost:8080/admin*, and you should now get the error RBAC: access denied, as shown in Figure 6-19.

Figure 6-19. The /admin page is now denied

You're getting this error because the intention is now blocking all requests to /admin from the ingress gateway. If you open up *http://localhost:8080*, you should see that regular requests are working without issue. Only the /admin requests are being blocked.

The next section describes how to configure application aware intentions on VMs. Feel free to skip ahead to "Application aware intentions in the Consul UI" on page 127.

Configuring application aware intentions on VMs

Currently, in *frontend-service-intentions.hcl* under the Sources list, you're specifying the service name and action, in this case, ingress-gateway and allow:

```
...
Sources = [
  {
    Name = "ingress-gateway"
    Action = "allow"
  }
]
```

Instead of setting the action to allow or deny, Consul supports a Permissions key that allows you to set specific HTTP-based permissions.

Edit *frontend-service-intentions.hcl* and change its contents to match Example 6-8.

Example 6-8. frontend-service-intentions.hcl

```
Kind = "service-intentions"
Name = "frontend"
Sources = [
  {
    Name = "ingress-gateway"
    Permissions = [ ❶
      {
```

```
      HTTP {
        PathPrefix = "/admin" ❷
      }
      Action = "deny"
    },
    {
      HTTP {
        PathPrefix = "/" ❸
      }
      Action = "allow"
    }
  ]
 }
]
```

❶   Consul supports a `Permissions` array in place of the `Action` key for specifying
more granular permissions. The first matching permission will apply.

❷   The first permission matches on all requests with the path prefix `/admin` and sets
the action to `deny`.

❸   The second permission matches all requests using the path prefix `/`. This permission will apply to any request that wasn't matched by the first permission and will
allow the request through.

 The full schema of the service intentions resource can be found in
Consul's reference documentation (*https://oreil.ly/WnK9i*).

Apply the change:

```
$ consul config write frontend-service-intentions.hcl
```

```
Config entry written: service-intentions/frontend
```

Now that the intention is applied, a regular request to the root (*/*) should succeed, but
a request with the prefix /admin, such as GET /admin, should fail.

To test this out, refresh *http://localhost:8080/admin* twice. You should see an RBAC:
access denied error as shown in Figure 6-20 on the second refresh.

Figure 6-20. The /admin page is now denied

Application aware intentions in the Consul UI

Now that you've confirmed that your application aware intention is working as expected, you can also view the intention in Consul's UI.

On the Intentions page (see Figure 6-21), the intention from the ingress gateway to frontend is now listed as "App aware."

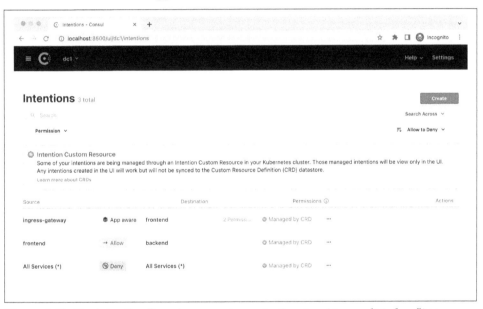

Figure 6-21. The intention from ingress gateway to frontend is now listed as "App aware"

If you click on that intention, you'll see the list of permissions at the bottom of the page as shown in Figure 6-22. The P stands for PathPrefix, which is the type of matching rule you configured.

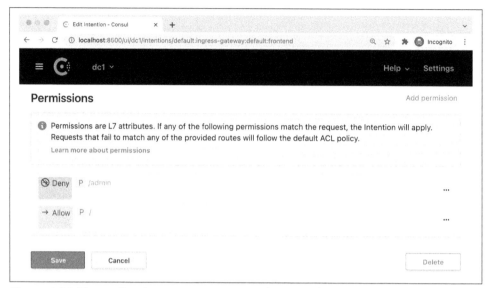

Figure 6-22. The application aware intention is shown in the UI with the matching rules

Summary

In this chapter, you learned about the security improvements you can implement using Consul service mesh. You saw how zero trust networks are more secure than traditional castle and moat networks, and I covered the three requirements of a zero trust network: encryption, authentication, and authorization.

The chapter discussed how TLS encryption works and how Consul uses mTLS to implement authentication and authorization. You then learned how to configure Consul's authorization rules, known as intentions.

You secured your example cluster so that only the frontend service could talk to the backend service, and only the ingress gateway could talk to the frontend service. You then further locked down permissions by preventing the ingress gateway from accessing the frontend service's /admin endpoints.

In the next chapter, you'll learn about service mesh observability—how to use the service mesh to understand the status of your services while they're running.

## CHAPTER 7

Observability

Observability is the ability to understand the state of a running system as an outside observer. Good observability is crucial in software systems because problems that occur in production often can't be reproduced in development environments.

When a system with good observability has an issue, engineers can quickly investigate and figure out what's happening by looking at the observability data. For example, if a service seems to be responding slowly, they may be able to look at the metrics and see that the machine the service is running on is using 100% of its CPU. Or, if users see a blank page when loading a frontend service, engineers might be able to look at the service's logs and see that it's erroring because it can't connect to its database.

In contrast, a system with poor observability is difficult to debug. If users are reporting errors that can't be reproduced in a development environment, then the engineers are left guessing as to the cause.

An observable system exposes three key types of telemetry:[1]

Metrics
> Statistics about the service or the underlying infrastructure. For example, how many requests per second the service is receiving or the current CPU and memory usage. Metrics are often viewed in aggregate—for example, the average number of requests per second across all instances of a particular service.

Logs
> Messages written by service developers that are output in response to certain events that occur in the service. For example, if a service can't connect with

---

1 *Telemetry* is a term used in software engineering to describe data about a system.

its database, it might log an error stating: `Cannot connect to database: 127.0.0.1:3306 connection refused`.

Logs are not discussed in detail in this chapter; you can configure sidecar proxies to log every request, but this results in too many logs to be useful in most cases. Instead, this information is best collected via metrics, which can be aggregated, and traces (see the next list item), which can be sampled.

Traces (or distributed traces)
Provide data about a single action or request that spans multiple microservices. For example, when you load a new bird in the Birdwatcher application, your browser makes a request to the ingress gateway. The ingress gateway then makes a request to the `frontend` service, which in turn makes a request to the `backend` service. A trace of that request will include data about the calls to each service along the request path.

In this chapter, you'll learn how Consul can improve observability in your systems. You'll deploy two open source tools for viewing metrics, Prometheus and Grafana, and configure Consul to emit metrics from its sidecar proxies. Later in the chapter, you'll learn how to instrument your services for distributed tracing, and you'll deploy Jaeger, an open source tracing tool, to view your traces.

Metrics

Metrics are helpful from an observability standpoint for many reasons:

- You can use them to diagnose why something is failing. For example, if a service is responding slowly, metrics might show that one of its dependencies has increased latency.

- You can use them to alert on potential issues before they become outages. For example, if you know that your service starts to degrade when it receives more than 1,000 requests per second (RPS), you could alert when it's receiving 750 RPS.

- You can use them to build dashboards. Dashboards are useful for a quick glance to understand a service's health.

- You can use them to view trends over time. For example, if a service is showing steady latency increases, you know you need to investigate further before it becomes a major issue.

Traditionally, metrics are obtained via two methods:

- A metrics agent runs on each machine and captures system metrics like CPU usage and RAM, or an agent runs in a Kubernetes cluster and captures cluster metrics from the Kubernetes API.
- Service developers write code to capture and emit metrics. For example, they will import a metrics library and increment a counter on every outgoing request.

The problem with the second method is that it relies on service developers to write extra code. For example, consider the following code:

```go
func makeRequest() error {
    err := http.Get("http://www.google.com")
    if err != nil {
        return err
    }
    return nil
}
```

If you wanted to instrument it with metrics, you would add something like the following:

```go
func makeRequest() error {
    start := time.Now() ❶

    err := http.Get("http://www.google.com")
    if err != nil {
        failureCounter.Increment() ❷
        return err
    }
    requestTimer.Record(time.Since(start)) ❸
    successCounter.Increment() ❹

    return nil
}
```

❶ Record the time before the request is made. This will be used to calculate how long the request took.

❷ Increment a counter if the request fails.

❸ Record how long the request took.

❹ Increment a counter if the request succeeds.

Adding instrumentation code by hand increases development time and is error-prone. It's easy for developers to forget to add instrumentation code in some places. In addition, the metrics that are produced might be inconsistent between services. This means operators will need to understand each service's unique metrics when diagnosing issues.

When faced with the problem of needing to ensure consistent metrics across services, many developers decide to write a shared library that is imported by all services. However, this has its own problems:

- You need to write a library for each language used.
- The library needs to work for all network protocols in use—for example, TCP, HTTP, and gRPC.
- You need to update all services to use the shared library, and it's time-consuming to modify old services and redeploy them. Some services might not be on an up-to-date version of the language or be using old dependencies that don't work with the shared library.
- If you decide to expose more metrics—for example, you forgot to capture TCP connection latency—you now need to update the library, update all services to use the latest version, and redeploy all services.

In short, shared libraries couple teams and services together, negating one of the main benefits of microservices.

A service mesh solves this problem in a better way. Since all calls go through the service mesh sidecars, the sidecars can automatically capture and emit detailed metrics (see Figure 7-1).

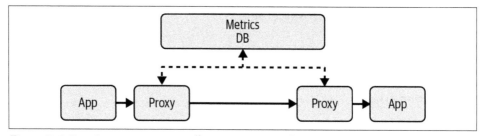

Figure 7-1. Proxies can automatically capture network metrics and emit them to a metrics database

Without any additional code or any change to your services, you can capture metrics for all incoming and outgoing network calls across every service in your infrastructure. This capability makes the service mesh a powerful tool for observability.

When you configure the service mesh to emit metrics, you need a place to store those metrics. There are many metrics storage solutions. Some are free and open source, such as Prometheus (*https://oreil.ly/fjkjW*), Graphite (*https://oreil.ly/mUznv*), Elasticsearch (*https://oreil.ly/HdOc4*), and InfluxDB (*https://oreil.ly/FXThN*), and some are paid solutions, such as Datadog (*https://oreil.ly/dBSZE*) and New Relic (*https://oreil.ly/uCzw8*).

This book uses Prometheus because it's currently the most popular open source metrics database. However, you can hook Consul up with any metrics storage that supports Prometheus metrics (which most do).

In the next section, you'll learn how to deploy and configure Prometheus.

Deploying and Configuring Prometheus

If you're using Kubernetes, Prometheus was already installed and configured back in Chapter 3 when you installed Consul. Prometheus was installed because your *values.yaml* file contained the following stanza:

```
prometheus:
  enabled: true
```

The Prometheus installation installed alongside Consul is only for development and testing purposes. If you want to install Prometheus for production, you should use the Prometheus Operator (*https://oreil.ly/cwOuA*) project.

Since Prometheus is already installed and configured on Kubernetes, if you're using Kubernetes, skip ahead to "Viewing Consul UI Metrics" on page 138. The rest of this section describes how to run Prometheus on VMs.

Prometheus is also already installed on the Vagrant VM, but it is not yet configured.

To install Prometheus on your own VMs, follow Prometheus's installation instructions (*https://oreil.ly/aBUcY*).

Prometheus is a pull-based metrics system, which means you must configure it with a list of endpoints to pull metrics from. Prometheus will then periodically make HTTP requests to these endpoints and store the metrics it receives. Prometheus calls its metric pulling operation *scraping*.

You want Prometheus to scrape metrics from each Consul sidecar, but it would be a big headache to specify the endpoint for every single sidecar. Luckily, Prometheus integrates with Consul and will query Consul for the list of addresses. It will then use those addresses to scrape metrics.

To configure Prometheus, first SSH into the VM:

```
$ vagrant ssh
```

Create a Prometheus config file at */etc/prometheus/prometheus.yml* with the contents from Example 7-1 (you will need `sudo`).

Example 7-1. /etc/prometheus/prometheus.yml

```
global:
  scrape_interval: 10s ❶

scrape_configs: ❷
- job_name: consul
  metrics_path: /metrics ❸
  consul_sd_configs: ❹
  - server: 'localhost:8500'
  relabel_configs: ❺
  - source_labels:
    - __meta_consul_tagged_address_lan
    - __meta_consul_service_metadata_prometheus_port
    regex: '(.*);(.*)'
    replacement: '${1}:${2}'
    target_label: '__address__'
    action: 'replace'
```

❶ The global scrape interval controls how often Prometheus scrapes for metrics.

❷ Under `scrape_configs`, the different jobs for scraping are configured. In this case there's a single job called `consul` that configures Prometheus to scrape Consul sidecar proxies.

❸ The HTTP path that proxies expose metrics on is `/metrics`.

❹ The `consul_sd_configs` stanza configures Prometheus to use Consul service discovery to discover the scraping addresses. It accepts the address of the Consul server.

❺ This section looks complicated but it's simply configuring the port that Prometheus uses to scrape for metrics. By default, Prometheus will try to use the service's port, but in this case it's the proxy exposing the metrics, not the service. This config instructs Prometheus to use the port set in the metadata key `prometheus_port`. You will then set this key to the proxy's Prometheus port in the service registration.

Change the ownership of */etc/prometheus/prometheus.yml* so Prometheus can read it:

```
$ sudo chown prometheus:prometheus /etc/prometheus/prometheus.yml
```

Now you're ready to start Prometheus:

```
$ sudo systemctl start prometheus
```

Check its status; it should be active and running:

```
$ sudo systemctl status prometheus

  prometheus.service - Monitoring system and time series database
    Active: active (running) since ...; 3s ago
```

You should now be able to view the Prometheus UI at *http://localhost:9090* (see Figure 7-2).

Figure 7-2. The Prometheus UI

Prometheus is now running, but the ingress gateway and sidecars are not yet exposing their metrics for Prometheus to scrape. In the next section, you'll configure your services and ingress gateway to expose their metrics.

Emitting Metrics

Consul sidecar proxies and ingress gateways don't expose Prometheus metrics by default. On VMs, you need to modify your service and ingress gateway registration files to enable metrics.

Edit */etc/consul.d/frontend.hcl* (using `sudo`) to match Example 7-2.

Example 7-2. /etc/consul.d/frontend.hcl

```
service {
  name = "frontend"
  port = 6060

  meta {
    prometheus_port = "20200"  ❶
  }

  connect {
    sidecar_service {
      port = 21000
      proxy {
        upstreams = [
          {
            destination_name = "backend"
            local_bind_port  = 6001
          }
        ]
        config {
          envoy_prometheus_bind_addr = "0.0.0.0:20200"  ❷
        }
      }
    }
  }
}
```

❶ Add a `meta` stanza and set a metadata key named `prometheus_port` to the port Envoy will serve Prometheus metrics on. This metadata key is read by the Prometheus server so it knows which port to use to scrape metrics for this service instance.

❷ Add a `config` stanza under `proxy` to set the address and port that Envoy will serve Prometheus metrics on. Port 20200 is used for this exercise, but it can be any free port.

Edit */etc/consul.d/backend.hcl* (also using `sudo`) and make similar changes so it matches Example 7-3.

Example 7-3. /etc/consul.d/backend.hcl

```
service {
  name = "backend"
  port = 7000

  meta {
    version = "v1"
    prometheus_port = "20201"
```

```
      }
  connect {
    sidecar_service {
      port = 22000
      proxy {
        config {
          envoy_prometheus_bind_addr = "0.0.0.0:20201" ❶
        }
      }
    }
  }
}
```

❶ For the backend service, use port 20201 so it doesn't conflict with frontend's port 20200.

Also, edit */etc/consul.d/ingress-gateway.hcl* to match Example 7-4.

Example 7-4. /etc/consul.d/ingress-gateway.hcl

```
service {
  name   = "ingress-gateway"
  kind   = "ingress-gateway"
  port   = 20000

  meta {
    prometheus_port = "20202"
  }

  checks = [
    {
      name     = "ingress-gateway listening"
      tcp      = "localhost:20000"
      interval = "30s"
    }
  ]
  proxy {
    config {
      envoy_prometheus_bind_addr = "0.0.0.0:20202" ❶
    }
  }
}
```

❶ The ingress gateway will use port 20202 for Prometheus metrics.

Finally, configure Consul so that its UI will pull metrics from Prometheus. You will then be able to view the metrics in the Consul UI's topology view. Edit */etc/consul.d/ server.hcl* and modify the ui_config stanza, so it looks as follows:

```
ui_config {
  enabled         = true
  metrics_provider = "prometheus"
  metrics_proxy {
    base_url = "http://localhost:9090" ❶
  }
}
```

❶ Consul will use this URL (which routes to the Prometheus service on the VM) to make requests to Prometheus.

Now that you've changed the configuration, reload Consul so it picks up the changes:

```
$ consul reload

Configuration reload triggered
```

Envoy only reads the Prometheus configuration on startup, so you must restart the sidecar proxies and ingress gateway.

```
$ sudo systemctl restart frontend-sidecar-proxy
$ sudo systemctl restart backend-sidecar-proxy
$ sudo systemctl restart ingress-gateway
```

Congratulations! You've successfully configured Consul to emit metrics and configured Prometheus to scrape them. In the next section, you'll generate metrics and view them in Consul's UI and Grafana.

Viewing Consul UI Metrics

By this point, you should have Prometheus installed on either Kubernetes or VMs, and it should be scraping your sidecar proxies and ingress gateway and capturing their metrics. Now it's time to view those metrics.

The easiest way to view the metrics is through Consul's UI. Open up *http://local host:8500* (if you're on Kubernetes, ensure you're running `minikube tunnel` or `kubectl port-forward`) and click the `frontend` service. On VMs, you'll need to do a hard refresh of the page by holding the Shift key and clicking the browser's refresh icon. You should see a page similar to Figure 7-3.

If you haven't made any requests to the Birdwatcher app in the last 15 minutes, the UI will show no metrics. Open up a new tab, navigate to *http://localhost:8080*, and toggle Auto Shuffle as shown in Figure 7-4.

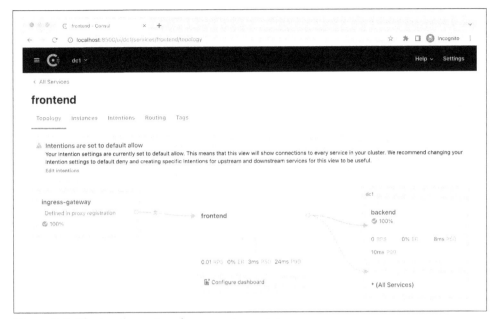

Figure 7-3. The topology view shows service metrics

Figure 7-4. Set the Birdwatcher app to Auto Shuffle to continuously generate metrics

This will generate a continuous stream of metrics by causing the `frontend` service to request a new bird every two seconds.

Leave the Birdwatcher tab open and switch back to the Consul UI's tab. After a couple of seconds, you will see metrics like in Figure 7-5.

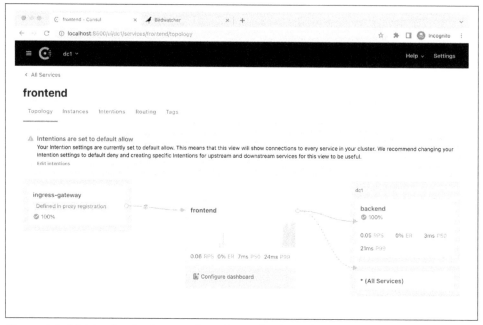

Figure 7-5. Metrics for the `frontend` service in the Consul UI

 If you don't see any metrics on Kubernetes, you may need to restart the prometheus server with: `kubectl rollout restart deploy/ prometheus-server -n consul`.[2]

The graph shows requests per second and errors per second of incoming requests. Currently, there are no errors, but they would be shown in red.[3] Along the bottom of the graph, there are four metrics shown:

---

2 I've seen this occur on minikube when Docker temporarily suspends the minikube container, which causes issues with time tracking within the Prometheus server.

3 If you want to see errors, increase the error rate in the Birdwatcher UI and then look at backend's metrics.

RPS

Requests per second. How many requests the `frontend` service is *receiving* per second.

ER

Error rate. What percentage of requests resulted in a 5xx HTTP error, for example, a 500 or a 503.

P50

Latency at the 50th percentile. In other words, 50% of requests were completed in this time or less.

P99

Latency at the 99th percentile. That is, 99% of requests were completed in this time or less.

> All the metrics are averaged over 15 minutes, which also matches the graph period. This means at the beginning, even though there is one request every two seconds to the `frontend` service, the average requests per second will be lower, since for the first number of minutes there were zero requests.

In addition to these four metrics, there are hundreds more metrics exposed by the sidecar proxies—for example, the number of active connections or how long TCP connections are taking to be established. Consul's UI just shows these four metrics because they provide the best insight into the health of the service without cluttering the UI.

> A list of all the exposed metrics is available through Envoy's documentation (*https://oreil.ly/qG61o*).

The metrics shown in the Consul UI are helpful for understanding the current state of your services, but inevitably you will need to see other metrics or metrics from different time periods than the last 15 minutes. That's where Grafana comes in.

Grafana

Grafana (*https://grafana.com*) is the leading open source technology for viewing metrics. Grafana is easily deployed and integrates seamlessly with Prometheus. In this section, you'll learn how to install Grafana and build Grafana dashboards.

Grafana is installed differently depending on whether it's running on Kubernetes or VMs. First, installation on Kubernetes is covered, and then on VMs ("Installing Grafana on VMs" on page 143).

Installing Grafana on Kubernetes

You can install Grafana via its Helm chart. First, you will need to install Helm by following the instructions on Helm's website (*https://oreil.ly/PiAUD*). The exercises were tested with Helm v3.8.1.

Now you're ready to install Grafana. Create a Kubernetes Secret for the Grafana administrator's username and password. Since this is for testing purposes, use *admin* and *admin*:

```
$ kubectl create secret generic grafana-admin \
    --from-literal=admin-user=admin \
    --from-literal=admin-password=admin
```

Use helm install to install Grafana:

```
$ helm install grafana grafana \
    --version 6.17.1 \
    --repo https://grafana.github.io/helm-charts \
    --set service.type=LoadBalancer \
    --set service.port=3000 \
    --set persistence.enabled=true \
    --set rbac.pspEnabled=false \
    --set admin.existingSecret=grafana-admin \
    --wait
```

The install should take 30 seconds to a minute to complete.

 If you're not running Kubernetes in minikube, or you're running minikube on Linux, set the service.type to ClusterIP. Otherwise, you may expose your Grafana server to the public internet. To view the Grafana UI, use kubectl port-forward service/grafana 3000.

Navigate to *http://localhost:3000* using your browser (be sure that minikube tunnel or kubectl port-forward is running), and you should see the Grafana login screen as shown in Figure 7-6.

Figure 7-6. Log in to Grafana with the username admin and the password admin

Log in using username *admin* and password *admin*. When prompted to set a new password, use *admin* again, for simplicity.

Congratulations, you've successfully installed Grafana! Skip ahead to "Configuring the Prometheus data source" on page 144.

Installing Grafana on VMs

Grafana is already installed on your Vagrant VM and configured via `systemd`.

> To install Grafana on your own VMs, see Grafana's installation instructions (*https://oreil.ly/5h1DG*).

Start Grafana using `systemctl`:

```
$ sudo systemctl start grafana-server
```

Navigate to *http://localhost:3000* using your browser, and you should see the Grafana login screen as shown in Figure 7-7.

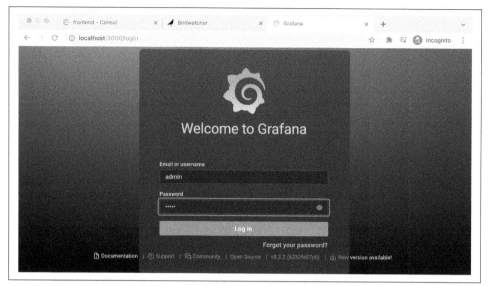

Figure 7-7. Log in to Grafana with the username admin and the password admin

Log in using the username *admin* and the default password *admin*. Use *admin* and *admin* on the next screen to permanently set that as the username and password.

Congratulations, you're running Grafana!

Configuring the Prometheus data source

Now that Grafana is running, you need to configure it to pull data from Prometheus. Grafana calls the metrics databases it pulls from *data sources*. To add Prometheus as a data source, perform the following steps:

1. Click the gear icon on the left-hand side navigation and select "Data sources."
2. Click "Add data source."
3. Select Prometheus.
4. Under the HTTP section in the URL field, fill in `http://prometheus-server.consul` if you're on Kubernetes or `http://localhost:9090` for VMs (see Figure 7-8).
5. Leave the rest of the configuration as is.
6. Scroll to the bottom and click "Save & test."
7. You should see a message that the data source is working.

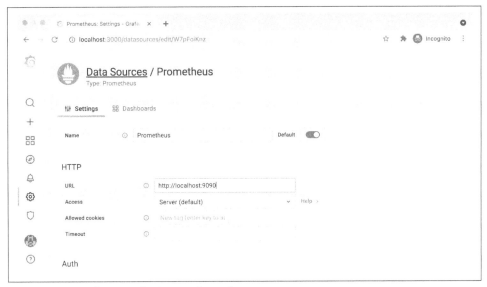

Figure 7-8. Configuring the Prometheus data source

Grafana is now configured to pull data from Prometheus, and you're ready to build your first Grafana dashboard.

Building Grafana dashboards

You're going to build a dashboard that uses the same metrics as the Consul UI. You can then add your own metrics once you understand how dashboards are constructed.

On the left-hand side navigation, click the plus icon and select Dashboard. This will bring you to the dashboard creation screen as shown in Figure 7-9.

Click "Add an empty panel," and you'll arrive at the panel creation page. Here you can enter queries against your data sources, and the results will be displayed on the graph.

There is a section on this page that prompts you to enter a query in Prometheus Query Language (PromQL) (see Figure 7-10).

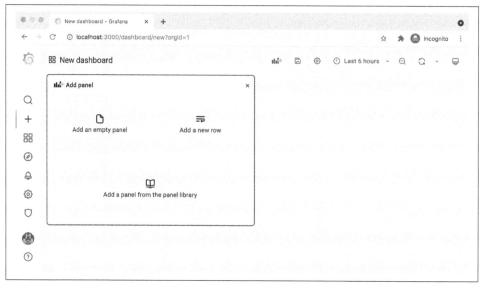

Figure 7-9. The Grafana dashboard creation screen

Figure 7-10. You can write PromQL queries into the "Metrics browser" field

PromQL enables you to write queries to display Prometheus metrics. The "Metrics browser" field is where you specify the query you want to show on the panel.

The first metric to put on your dashboard is the number of requests per second (RPS) received by the `frontend` service.

Every Envoy proxy records the number of incoming requests using the counter `envoy_http_downstream_rq_completed`. The parts of the metric name correspond to the following:

envoy
> The prefix for all metrics that come from the Envoy sidecars.

http
> Means the metric corresponds to HTTP requests.

downstream
> Means the metric is counting requests that are *received* by a service. Upstream requests are requests that are *sent* to another service.

rq
> Stands for request.

completed
> Means that this metric counts the number of completed HTTP requests. Some requests might be aborted before they can be completed, so those would not be counted here.

You can't just enter that metric name as the PromQL query to show RPS for the frontend service, however. Instead, you need to use the query listed in Example 7-5.

Example 7-5. PromQL query for requests per second to the frontend service

```
sum( ❶
  rate( ❷
    envoy_http_downstream_rq_completed{ ❸
      consul_source_service="frontend", ❹
      envoy_http_conn_manager_prefix="public_listener" ❺
    }[$__rate_interval] ❻
  )
)
```

❶ Sum the requests across all instances of the frontend service. You're only running one instance right now, but if you scale this up later, the metric will show the total RPS across all instances.

❷ You want to show the requests *per second*, which is a rate. Otherwise, your graph will be a line showing total requests.

❸ The metric to display.

❹ Narrow down the metric to the one emitted from the frontend service. Remember that every Envoy proxy emits this metric.

❺ Further narrow down the metric to the Envoy listener that handles incoming traffic. Consul sidecar proxies have other listeners, such as for their admin portal, and you don't want to count those requests as part of this query.

❻ This value controls over what period the rate is averaged. $\_\_rate\_interval$ is a Grafana variable that Grafana will set to the appropriate period depending on the time horizon of the graph.

Press Shift+Enter, and change the time period from 6 hours to the last 15 minutes. You should see the RPS show up on the graph (see Figure 7-11).

Figure 7-11. After entering the query, you see the RPS graph

Your metric and graph are now fully specified. Change the panel title (on the right side) to "Requests per second," and then click Apply at the top right to add this panel to your dashboard. Your dashboard should look like Figure 7-12.

 Grafana panels are highly customizable. You can also experiment with changing the line colors, axis labels, and more.

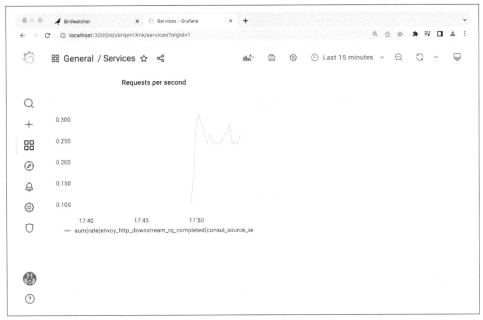

Figure 7-12. The Grafana dashboard after adding the first panel

Before you add more panels, save the dashboard by clicking the Save icon in the top right. Name the dashboard "Services" and use the default folder "General."

Currently, this dashboard only shows metrics for the `frontend` service. If you want to reuse the same dashboard for multiple services so you don't have to re-create the same panels, you can use *Grafana variables*.

To configure a dashboard's variables, click the Dashboard settings gear icon at the top right (not the gear icon in the left sidebar). In the settings screen for the dashboard, select the Variables menu on the left.

Click "Add variable" and set the following fields (see Figure 7-13):

- *Name* should be set to `Service`
- *Query* under *Query options* should be set to:

      label_values(consul_source_service)

   This will make the variable a drop-down menu populated with a list of Consul service names.

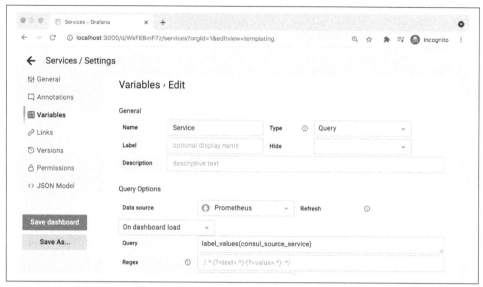

Figure 7-13. Add a variable that allows you to switch between different services

Click Update to save the variable, then click "Save dashboard" on the left. You can leave the note in the "Save dashboard" dialog blank, and click Save. Click the left-facing arrow at the top left of the page to go back to the dashboard.

Now you should see a drop-down menu that lets you select between the service names frontend, backend, and ingress-gateway (as shown in Figure 7-14).

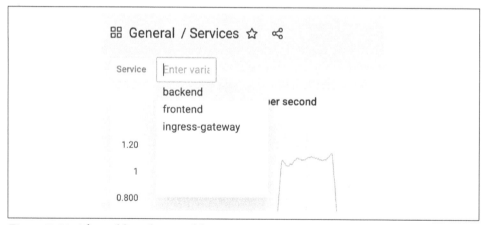

Figure 7-14. After adding the variable, you can now switch between services

Currently, changing the service does nothing. This is because the panel's query is hardcoded to always use the frontend service. To modify the panel to use the variable, click the panel's title and select Edit. Variables are referenced using the syntax $variable_name, so change the query to the following:

```
sum(
  rate(
    envoy_http_downstream_rq_completed{
      consul_source_service="$Service",
      envoy_http_conn_manager_prefix="public_listener"
    }[$__rate_interval]
  )
)
```

Apply the change by clicking Apply.

Now when you change the variable between the frontend and backend services, the graph should change slightly. It won't be a significant change because both services are receiving the same requests per second.

Although ingress-gateway is listed as a service name, it won't display any metrics. This is because ingress gateways use a different listener name than regular sidecar proxies. They use the name ingress_upstream_8080 instead of public_listener.[4]

Edit the panel again and modify your query to match on either listener name:

```
sum(
  rate(
    envoy_http_downstream_rq_completed{
      consul_source_service="$Service",
      envoy_http_conn_manager_prefix=~"public_listener|ingress_upstream_8080"
    }[$__rate_interval]
  )
)
```

> Note that envoy_http_conn_manager_prefix=... changed to envoy_http_conn_manager_prefix=~... (= to =~) to allow matching on either listener name.

Apply the change, and now when you switch the variable to the ingress gateway, you should see metrics.

Now that you've configured your dashboard variables and first panel, you can add additional panels to show more metrics. Click the "Add panel" icon at the top right to add new panels.

---

4 The names of ingress gateway listeners are suffixed with their port numbers.

Add a panel named "Error %" using the query from Example 7-6. Currently there should be no errors, so there will be a line at 0.

Example 7-6. PromQL query showing percent of requests that resulted in an error

```
sum(
  rate(
    envoy_http_downstream_rq_xx{
      consul_source_service="$Service",
      envoy_http_conn_manager_prefix=~"public_listener|ingress_upstream_8080",
      envoy_response_code_class="5"
    }[$__rate_interval]
  )
) /
sum(
  rate(
    envoy_http_downstream_rq_completed{
      consul_source_service="$Service",
      envoy_http_conn_manager_prefix=~"public_listener|ingress_upstream_8080"
    }[$__rate_interval]
  )
)
```

Finally, add a panel named "Latency" that uses two queries, the first for 50th percentile latency (Example 7-7) and the second for 99th percentile latency (Example 7-8).

Example 7-7. PromQL query calculating 50th percentile latency

```
histogram_quantile(
  0.5,
  sum(
    rate(
      envoy_http_downstream_rq_time_bucket{
        consul_source_service="$Service",
        envoy_http_conn_manager_prefix=~"public_listener|ingress_upstream_8080"
      }[$__rate_interval]
    )
  ) by (le)
)
```

Example 7-8. PromQL query calculating 99th percentile latency

```
histogram_quantile(
  0.99,
  sum(
    rate(
      envoy_http_downstream_rq_time_bucket{
        consul_source_service="$Service",
        envoy_http_conn_manager_prefix=~"public_listener|ingress_upstream_8080"
```

```
      }[$__rate_interval]
    )
  ) by (le)
)
```

 Add multiple queries to the same panel by clicking the "+ Query" button at the bottom of the panel editing view.

When complete, your dashboard should look something like Figure 7-15. Feel free to continue to tweak your dashboard and add new panels. You can review Envoy's stats documentation (*https://oreil.ly/c6Fp6*) to see the other metrics it emits.

When you're satisfied, don't forget to save the dashboard.

Figure 7-15. The finished dashboard

Once you've configured your dashboard to your liking, the next step is to link to it from within the Consul UI.

Linking to dashboards from the Consul UI

You can configure the Consul UI to display links to metrics dashboards like Grafana, as shown in Figure 7-16. This makes it easy for users to dive deeper into the metrics for a specific service.

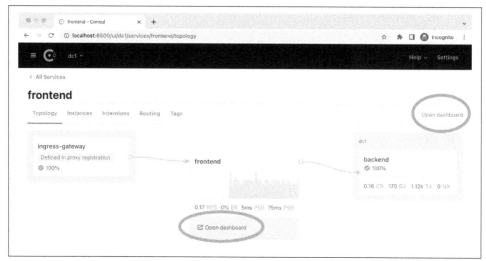

Figure 7-16. You can configure the Consul UI to display links to metrics dashboards

To enable this feature, you configure Consul with a templated URL to your metrics dashboard, like `http://your.dashboard/{{Service.Name}}`. The `{{Service.Name}}` section will be replaced with the name of the service the user is viewing.

The first step to configuring Consul is coming up with your templated URL. Navigate to your dashboard and look at the URL; it should be something like:

```
http://localhost:3000/d/WkFEBmF7z/services?orgId=1
```

 The `WkFEBmF7z` part will be different for your dashboard.

This URL links to your dashboard, but it will always show the metrics of the same service. Luckily, Grafana supports setting the Service variable in the URL by appending:

```
&var-Service=<service>
```

You can use this to generate your templated URL:

```
http://localhost:3000/d/<your-id>/services?orgId=1&var-Service={{Service.Name}}
```

Now that you know your templated URL, you're ready to configure Consul. On Kubernetes, modify your *values.yaml* file's ui stanza, so it looks like this:

```
ui:
  service:
    type: LoadBalancer
    port:
      http: 8500
  dashboardURLTemplates:
    service: "http://localhost:3000/d/<your-id>/\  ❶
      services?orgId=1&var-Service={{Service.Name}}"
```

❶ Be sure to replace <your-id>.

Once you've modified your *values.yaml* file, run consul-k8s upgrade to update your servers:

```
$ consul-k8s upgrade -config-file values.yaml
```

On VMs, edit */etc/consul.d/server.hcl* and modify the ui_config stanza so it looks as follows:

```
ui_config {
  enabled          = true
  metrics_provider = "prometheus"
  metrics_proxy {
    base_url = "http://localhost:9090"
  }
  dashboard_url_templates {
    service =
      "http://localhost:3000/d/<your-id>/services?
        orgId=1&var-Service={{Service.Name}}" ❶
  }
}
```

❶ Be sure to use your own URL here. This line is split so it fits on the page, but it should be a single line in your own config. Also, be sure to replace <your-id>.

Then run consul reload so Consul picks up the new config:

```
$ consul reload
```

Once you've reconfigured Consul on Kubernetes or VMs, you will need to do a hard refresh of the Consul UI on *http://localhost:8500* to clear the browser's cache.

You can perform a hard refresh in most browsers by holding down the Shift key and pressing the refresh icon.

If everything was set up successfully, you should see links to your dashboard in the topology views! Navigate to each service and click on "Open dashboard" to test out the links.

Congratulations. You've successfully configured Consul to emit metrics to Prometheus, and built your first Grafana dashboard. Every service in the mesh is now instrumented with detailed request metrics that greatly increase the observability of your system; all without any code changes.

In the next section, you'll learn about another type of telemetry: distributed tracing.

Distributed Tracing

Distributed tracing provides observability for requests that span microservices. For example, a single request to the ingress gateway results in a call to the frontend service, which then calls the backend service. Distributed tracing correlates and collects these calls into a single trace. Operators can view these traces and see all of the microservices calls involved in responding to a single request.

Traces are extremely useful for investigating failures in microservices environments. For example, imagine that you're getting reports that users are seeing HTTP 500 errors in the Birdwatcher UI. The 500 errors could result from a failure anywhere in the request path: it could be the ingress gateway, the frontend service itself, or the backend service. You could look at the individual logs and metrics from each service and try to find the error yourself, but a trace will show you all of the calls involved and exactly which one failed.

How Tracing Works

The first step in recording a trace is generating a unique request ID that will be used to identify the request across all the microservices involved. The request ID is passed along as a header to each service. If a service receives a request without a request ID header, it will assume it's the first service in the request path and will generate the request ID itself.

In addition to passing along the request ID, each microservice the request passes through will also generate a span. A *span* is a defined segment of a trace. The services will emit their spans to the tracing collector, as shown in Figure 7-17, which will then use the request ID to correlate all the spans into a single trace.

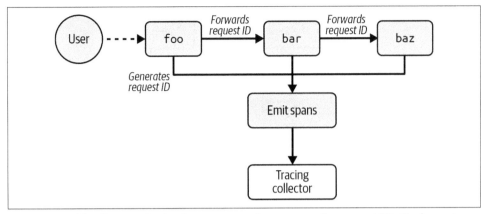

Figure 7-17. The first service in the request path generates the request ID. Each service passes along the ID to the next service in the path and emits its span to the tracing collector.

A span contains the following information:

- The service it's being emitted from
- The request ID
- The time the span started and when it ended
- Additional information, such as HTTP codes, errors, and logs

> Services are free to generate spans within themselves as well. For example, if a service must do a long computation, it might create a span for that computation. That span can then be viewed in the trace and used to understand why the request took as long as it did.

Instrumenting Your Services

Unlike other service mesh features such as encryption and metrics, service code must be modified to be distributed tracing aware. Services must ensure that they pass along the request ID header to subsequent requests because service mesh proxies can't correlate which incoming request resulted in which outgoing request.

For example, imagine if two requests, Request A and Request B, enter service `foo`, as shown in Figure 7-18. Service `foo` then makes two subsequent calls to service `bar` and `baz`. Without a request ID, `foo`'s proxy can't know which call resulted from Request A or B. In fact, maybe Request B resulted in no calls and Request A resulted in both.

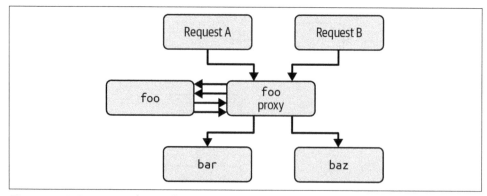

Figure 7-18. Proxies can't correlate requests without the service propagating request IDs

To instrument your services, you must decide on a tracing library. A tracing library handles emitting spans to the tracing collector and provides helper functions for propagating request IDs to upstream services. The most popular tracing library today is OpenTelemetry (*https://oreil.ly/GasCW*). Unfortunately, Envoy does not yet support adding its own spans to OpenTelemetry traces.[5] This means if you use OpenTelemetry for your services, you won't see spans from sidecar proxies and ingress gateways. Instead, you'll only see the spans emitted by your services. For this reason, the Birdwatcher application is configured to emit spans using an older format called Zipkin (*https://zipkin.io*), which is supported by Envoy. Zipkin has largely the same functionality as OpenTelemetry, but is older and does not have as active an open source community.

After you've decided on a tracing library and followed its documentation to instrument your services, the next step is to install a tracing collector.

Tracing Collectors

Tracing collectors collect the spans emitted by services and sidecar proxies and correlate them into a single trace. They also provide a UI for searching and viewing traces. The most popular open source tracing collector today is Jaeger (*https://oreil.ly/xdAAr*).[6] If you're on VMs, skip ahead to "Installing Jaeger on VMs" on page 162; otherwise, read on to learn how to install Jaeger on Kubernetes.

---

5 See the GitHub issue "Tracing: Transition to OpenTelemetry from OpenTracing and OpenCensus" (*https://oreil.ly/dqdgP*) for more information.

6 There are also many hosted collectors such as Datadog (*https://www.datadoghq.com*), Honeycomb (*https://www.honeycomb.io*), and Lightstep (*https://lightstep.com*).

Installing Jaeger on Kubernetes

Install the Jaeger operator using Helm:

```
$ helm install jaeger jaeger-operator \
    --version 2.26.0 \
    --repo https://jaegertracing.github.io/helm-charts \
    --wait
```

An *operator* is a Kubernetes deployment pattern where instead of directly installing specific software, you install its operator and then instruct the operator to install the software itself. This allows the operator to orchestrate complicated install and upgrade scenarios.

Next, create a resource that instructs the Jaeger operator to install the Jaeger collector using a demo configuration. Create a file *jaeger.yaml* in *manifests/* with the contents from Example 7-9.

Example 7-9. jaeger.yaml

```
apiVersion: jaegertracing.io/v1
kind: Jaeger
metadata:
  name: jaeger
spec:
  query:
    serviceType: LoadBalancer
  ingress:
    enabled: false
```

If you're not running Kubernetes in minikube, or you're running minikube on Linux, change the serviceType to ClusterIP and use kubectl port-forward service/jaeger-query 16686 to access the Jaeger UI. Otherwise, you may expose Jaeger to the public internet.

Apply the file to the cluster:

```
$ kubectl apply -f jaeger.yaml
```

```
jaeger.jaegertracing.io/jaeger created
```

Wait for the Jaeger collector to come up:

```
$ kubectl rollout status --watch deploy/jaeger
```

```
deployment "jaeger" successfully rolled out
```

To check that Jaeger was installed successfully, ensure that `minikube tunnel` is running (or `kubectl port-forward`) and then navigate to *http://localhost:16686*. You should see the Jaeger UI as shown in Figure 7-19.

Figure 7-19. The Jaeger UI

Now that Jaeger is installed, you're ready to configure your services to emit traces. The following section walks through how to do so using the example `frontend` and `backend` services.

Emitting service traces on Kubernetes

The `frontend` and `backend` services have built-in support for tracing that can be enabled by setting the environment variable `TRACING_URL`. You need to set the variable to the URL of the collector, which is `http://jaeger-collector.default:9411`.

Modify *frontend-deployment.yaml* to add the `TRACING_URL` environment variable as shown in Example 7-10.

Example 7-10. frontend-deployment.yaml

```
apiVersion: apps/v1
kind: Deployment
metadata:
  name: frontend
  labels:
    app: frontend
spec:
  replicas: 1
```

```
  selector:
    matchLabels:
      app: frontend
  template:
    metadata:
      labels:
        app: frontend
      annotations:
        consul.hashicorp.com/connect-inject: "true"
    spec:
      containers:
        - name: frontend
          image: ghcr.io/consul-up/birdwatcher-frontend:1.0.0
          env:
            - name: BIND_ADDR
              value: "0.0.0.0:6060"
            - name: BACKEND_URL
              value: "http://backend"
            - name: TRACING_URL
              value: "http://jaeger-collector.default:9411"
          ports:
            - containerPort: 6060
```

Make the same change to *backend-deployment.yaml*, as shown in Example 7-11.

Example 7-11. backend-deployment.yaml

```
apiVersion: apps/v1
kind: Deployment
metadata:
  name: backend
  labels:
    app: backend
spec:
  replicas: 1
  selector:
    matchLabels:
      app: backend
  template:
    metadata:
      labels:
        app: backend
      annotations:
        consul.hashicorp.com/connect-inject: "true"
        consul.hashicorp.com/service-meta-version: "v1"
    spec:
      containers:
        - name: backend
          image: ghcr.io/consul-up/birdwatcher-backend:1.0.0
          env:
            - name: BIND_ADDR
```

```
        value: "0.0.0.0:7000"
      - name: TRACING_URL
        value: "http://jaeger-collector.default:9411"
    ports:
      - containerPort: 7000
```

Then reapply both deployments:

```
$ kubectl apply -f frontend-deployment.yaml -f backend-deployment.yaml
```

```
deployment.apps/frontend configured
deployment.apps/backend configured
```

Wait for both deployments to roll out:

```
$ kubectl rollout status --watch deploy/frontend
```

```
deployment "frontend" successfully rolled out
```

```
$ kubectl rollout status --watch deploy/backend
```

```
deployment "backend" successfully rolled out
```

Now that you've configured the services to emit traces, you're ready to analyze them through Jaeger's UI. Skip ahead to "Viewing Service Traces" on page 164.

Installing Jaeger on VMs

Jaeger is already installed on the Vagrant VM.[7] To start it, first SSH into the VM:

```
$ vagrant ssh
```

Now start Jaeger using systemctl:

```
$ sudo systemctl start jaeger
```

Check its status; it should be active and running:

```
$ sudo systemctl status jaeger
...
    Active: active (running) since ...; 3s ago
```

To check that Jaeger is working as expected, navigate to *http://localhost:16686*. You should see the Jaeger UI as shown in Figure 7-20.

---

7 You can install Jaeger on your own VMs by downloading it from its releases page (*https://oreil.ly/g3rBF*).

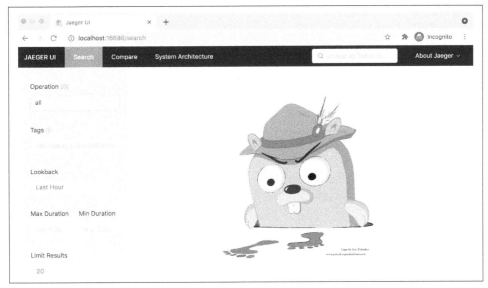

Figure 7-20. The Jaeger UI

Now that Jaeger is installed, you're ready to configure your services to emit traces.

Emitting service traces on VMs

The `frontend` and `backend` services already support tracing; they just need to be configured to emit their traces to Jaeger. To configure them, set the `TRACING_URL` environment variable to Jaeger's collection endpoint `http://localhost:9411`.

Edit */etc/systemd/system/frontend.service*, adding the following line:

```
Environment=TRACING_URL="http://localhost:9411"
```

The entire file should look like Example 7-12.

Example 7-12. /etc/systemd/system/frontend.service (with comments removed)

```
[Unit]
Description="Frontend service"
Requires=network-online.target
After=network-online.target

[Service]
ExecStart=/usr/local/bin/frontend
Restart=on-failure
Environment=BIND_ADDR=127.0.0.1:6060
Environment=BACKEND_URL=http://localhost:6001
Environment=TRACING_URL="http://localhost:9411" ❶
```

```
[Install]
WantedBy=multi-user.target
```

❶ The TRACING_URL environment variable is added here.

Next, edit */etc/systemd/system/backend.service* so it also sets that environment variable as shown in Example 7-13.

Example 7-13. /etc/systemd/system/backend.service (with comments removed)

```
[Unit]
Description="Backend service"
Requires=network-online.target
After=network-online.target

[Service]
ExecStart=/usr/local/bin/backend
Restart=on-failure
Environment=BIND_ADDR=0.0.0.0:7000
Environment=TRACING_URL="http://localhost:9411"

[Install]
WantedBy=multi-user.target
```

Reload systemd so it picks up the changes:

```
$ sudo systemctl daemon-reload
```

Now restart the two services:

```
$ sudo systemctl restart frontend backend
```

Congrats! Your services are now configured to emit traces to Jaeger!

Viewing Service Traces

Now that you've installed Jaeger and configured your services to emit traces, you're ready to analyze the trace data.

First, make a request to generate a trace by navigating to *http://localhost:8080*.

Once you've made that request, navigate to the Jaeger UI at *http://localhost:16686* to view the generated trace (if you already had the page open, refresh it). On the Search page, select frontend from the Service drop-down, and click Find Traces.

You should see a list of two or more traces, as shown in Figure 7-21.

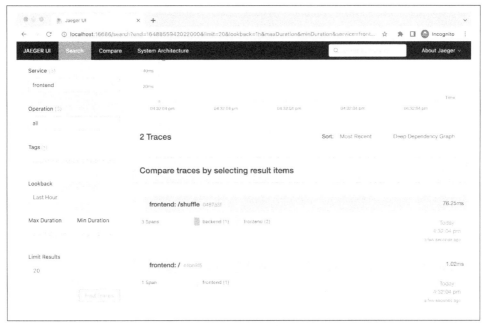

Figure 7-21. Two traces are created from loading the `frontend` *service*

The Jaeger UI lists traces from newest to oldest. The first trace in the list is a request to the `frontend` service on `/shuffle`. This trace comes from the request that the UI makes to load a bird. The request to `/shuffle` then triggers a subsequent call to the `backend` service so that trace involves two services. The next trace in the list is from loading the `frontend` UI at `/`. Loading the UI does not involve the `backend` service, so it's not part of that trace.

Click the `/shuffle` trace to navigate to the trace detail view as shown in Figure 7-22.

Figure 7-22. The trace detail view shows you data about a specific trace and all its spans

The trace detail view shows data about each span in the trace.

You should see three spans. Each span has a service and operation name. The two spans from the frontend service should have operation names /shuffle and call_backend. These identify the request from the UI to the frontend service on /shuffle and the subsequent request the frontend service made to the backend service. The span from the backend service is from receiving the /bird request from the frontend service.

Some other information to note here is that you can see the overall duration for the entire trace (in this example, it was 76.25 ms) and how many services were involved (2). You can also click into each span and see the tags attached to that specific span. The tags give you extra information, such as the HTTP status code and the request URL.

Congratulations! You've successfully enabled tracing in your services. In the next section, you'll look at also enabling tracing for the service mesh components.

Enabling Tracing for the Service Mesh

Currently you're seeing the spans from the frontend and backend services, but you're missing data about the request passing through the ingress gateway and the sidecar proxies. If those components added their own spans, you'd have a better picture of what's happening on each request (see Figure 7-23).

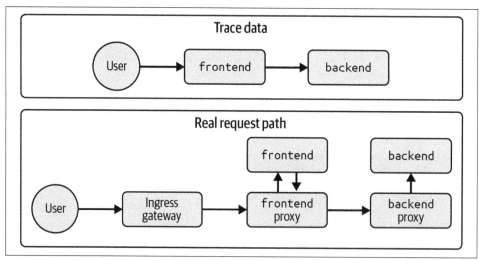

Figure 7-23. The current trace data only shows the requests between the frontend and backend services. In reality, the request is also passing through an ingress gateway and two sidecar proxies.

Luckily, Consul can configure Envoy to emit these spans. To do so, you need to modify the proxy defaults configuration that you first added in Chapter 5.

First, I cover configuring Kubernetes, and then VMs ("Service mesh tracing on VMs" on page 169).

Service mesh tracing on Kubernetes

You need to modify the `ProxyDefaults` resource to configure all proxies and ingress gateways to emit traces to Jaeger. Consul exposes Envoy's tracing configuration directly and allows you to pass in Envoy JSON config. This capability gives you more flexibility for configuring the exact settings for each tracing library and collector.

Edit *proxy-defaults.yaml* in *manifests/* and modify it to match Example 7-14.

Example 7-14. proxy-defaults.yaml

```
apiVersion: consul.hashicorp.com/v1alpha1
kind: ProxyDefaults
metadata:
  name: global
  namespace: consul
spec:
  config:
    protocol: http
    envoy_tracing_json: |  ❶
      {
        "http":{
          "name":"envoy.tracers.zipkin",
          "typedConfig":{
            "@type":"type.googleapis.com/envoy.config.trace.v3.ZipkinConfig",
            "collector_cluster":"jaeger_collector",
            "collector_endpoint_version":"HTTP_JSON",
            "collector_endpoint":"/api/v2/spans",
            "shared_span_context":false
          }
        }
      }
    envoy_extra_static_clusters_json: |  ❷
      {
        "connect_timeout":"3.000s",
        "dns_lookup_family":"V4_ONLY",
        "lb_policy":"ROUND_ROBIN",
        "load_assignment":{
          "cluster_name":"jaeger_collector",
          "endpoints":[
            {
              "lb_endpoints":[
                {
                  "endpoint":{
```

```
        "address":{
          "socket_address":{
            "address":"jaeger-collector.default",
            "port_value":9411,
            "protocol":"TCP"
          }
        }
      }
    }
  ]
}
],
},
"name":"jaeger_collector",
"type":"STRICT_DNS"
}
```

❶ envoy_tracing_json configures Envoy's tracing configuration. Here you're set-ting it to emit the traces to the collector named jaeger_collector.[8]

❷ envoy_extra_static_clusters_json looks complicated but it's just configuring the address of the Jaeger collector as jaeger-collector.default:9411 (the Kubernetes DNS address of Jaeger's Kubernetes service).

Apply the new ProxyDefaults config:

```
$ kubectl apply -f proxy-defaults.yaml
```

```
proxydefaults.consul.hashicorp.com/global configured
```

Ensure it is synced:

```
$ kubectl get proxydefaults global -n consul
```

```
NAME      SYNCED    LAST SYNCED    AGE
global    True      1s             1d
```

Now you must restart all of the Envoy sidecars and the ingress gateway. Most config-uration changes are picked up automatically by Envoy, but these particular changes require a restart because they're modifying configuration that is only read on startup.

Restart the ingress gateway, frontend and backend deployments using the kubectl rollout restart command:

```
$ kubectl rollout restart deploy/consul-ingress-gateway -n consul
$ kubectl rollout restart deploy/frontend
$ kubectl rollout restart deploy/backend
```

---

8 Jaeger has its own trace format, but it can also consume Zipkin-formatted traces.

```
deployment.apps/consul-ingress-gateway restarted
deployment.apps/frontend restarted
deployment.apps/backend restarted
```

Wait for the deployments to restart successfully:

```
$ kubectl rollout status deploy/consul-ingress-gateway --watch -n consul

deployment "consul-ingress-gateway" successfully rolled out

$ kubectl rollout status deploy/frontend --watch

deployment "frontend" successfully rolled out

$ kubectl rollout status deploy/backend --watch

deployment "backend" successfully rolled out
```

Consul is now configured to also emit trace data from ingress gateways and sidecar proxies! Skip ahead to "Analyzing Service Mesh Traces" on page 171 to learn how to view the new traces.

Service mesh tracing on VMs

You need to modify the proxy defaults config entry to configure all proxies and ingress gateways to emit traces to Jaeger. Consul exposes Envoy's tracing configuration directly and allows you to pass in Envoy JSON config. This capability gives you more flexibility for configuring the exact settings for each tracing library and collector.

vagrant ssh into the VM and edit *~/proxy-defaults.hcl* to match Example 7-15.

Example 7-15. proxy-defaults.hcl

```
Kind = "proxy-defaults"
Name = "global"
Config {
  protocol          = "http"
  envoy_tracing_json = <<EOF ❶
{
  "http":{
    "name":"envoy.tracers.zipkin",
    "typedConfig":{
      "@type":"type.googleapis.com/envoy.config.trace.v3.ZipkinConfig",
      "collector_cluster":"jaeger_collector",
      "collector_endpoint_version":"HTTP_JSON",
      "collector_endpoint":"/api/v2/spans",
      "shared_span_context":false
    }
  }
}
```

```
EOF
  envoy_extra_static_clusters_json = <<EOF ❷
{
  "connect_timeout":"3.000s",
  "dns_lookup_family":"V4_ONLY",
  "lb_policy":"ROUND_ROBIN",
  "load_assignment":{
    "cluster_name":"jaeger_collector",
    "endpoints":[
      {
        "lb_endpoints":[
          {
            "endpoint":{
              "address":{
                "socket_address":{
                  "address":"localhost",
                  "port_value":9411,
                  "protocol":"TCP"
                }
              }
            }
          }
        ]
      }
    ]
  },
  "name":"jaeger_collector",
  "type":"STRICT_DNS"
}
EOF
}
```

❶   envoy_tracing_json configures Envoy's tracing configuration. Here you're setting it to emit the traces to the collector named jaeger_collector.[9]

❷   envoy_extra_static_clusters_json looks complicated, but it's just configuring the address of the Jaeger collector as localhost:9411.

Write the updated config to Consul:

```
$ consul config write ~/proxy-defaults.hcl

Config entry written: proxy-defaults/global
```

---

9 As mentioned in the section "Service mesh tracing on Kubernetes" on page 167, Jaeger has its own trace format, but it can also consume Zipkin-formatted traces.

Now you must restart all of the Envoy sidecars and the ingress gateway. Envoy doesn't get automatically configured in this case because the tracing configuration is only picked up when Envoy first starts.

```
$ sudo systemctl restart ingress-gateway
$ sudo systemctl restart frontend-sidecar-proxy
$ sudo systemctl restart backend-sidecar-proxy
```

Consul is now configured to also emit trace data! In the next section, you'll generate more traces that now contain spans from the service mesh.

Analyzing Service Mesh Traces

At this point, you've enabled tracing for your services and the service mesh itself. To test it all out, make another request through the ingress gateway by loading the Birdwatcher UI at *http://localhost:8080*.

Now reload the Jaeger UI at *http://localhost:16686*, and you should be able to select the *ingress-gateway* service from the Service drop-down (you may need to refresh a couple of times because the traces aren't sent instantly). If you click Find Traces, you will now see a trace named *ingress-gateway: localhost:8080*. Click the trace to view its detailed view. It should look something like Figure 7-24.

Figure 7-24. The trace through the ingress gateway and sidecar proxies

This trace now lists three services and seven spans. Previously it was two services and three spans. The extra service is due to the ingress gateway being included in the trace, and the extra spans are due to the ingress gateway and the sidecar proxies adding their own spans.

It's easy to see the span added by the ingress gateway because its service name is `ingress-gateway`. The spans added by the sidecar proxies are harder to spot because they use the same service name as their services, but you can tell them apart by expanding the span details and looking for the spans tagged with `component = proxy`. Sidecar proxies add spans on incoming and outgoing requests, so you should see three spans from the sidecar proxies:

1. The incoming request from the ingress gateway to the `frontend` service. This will be listed under the `frontend` service.

2. The outgoing request from the `frontend` service to the `backend` service. This will also be listed under the `frontend` service since the outgoing request passes through `frontend`'s proxy.

3. The incoming request from the `frontend` service to the `backend` service. This will be listed under the `backend` service since the request passes through `backend`'s proxy after it leaves `frontend`'s proxy.

The spans added by the ingress gateway and sidecar proxies are useful for debugging issues that might be caused by the service mesh. For example, if you think a proxy is the cause of an unexpected latency spike, you can look at its span in the trace and see exactly how much latency it added.

This concludes the section on distributed tracing. In this section, you learned how distributed tracing works and how it can help debug issues in microservices environments. You then deployed Jaeger and configured the example services and Consul to emit traces to it.

Summary

In this chapter, you learned how Consul can increase the observability of your services. Without any modification to service code, you saw how the Envoy sidecar proxies can emit a full spectrum of metrics for each service. This capability can enable a small operations team to gain deep insight into every service in their infrastructure without any interaction with the teams that own the underlying services.

Next, you learned how to use Prometheus to store Envoy's metrics and Grafana to view them. You built a basic dashboard in Grafana using PromQL queries, and because the metrics were consistent across services, you were able to use the same dashboard for all services.

You then learned about distributed tracing and how it can help you track errors across microservices. You deployed Jaeger and learned how to configure your services and proxies to output spans to form a trace.

Increased observability helps track down problems more quickly, but what if you could prevent some issues from arising in the first place? That's the topic of the next chapter: reliability.

## CHAPTER 8

Reliability

Everything fails, all the time.

> —Werner Vogels, CTO of Amazon Web Services

In software systems, there's always something that can go wrong. Perhaps a database connection pool fills up, perhaps there's unexpected latency due to network issues, or maybe there's a bad deployment and a service starts responding with 500s. The likelihood of encountering failure is even higher in microservices infrastructures because a single request can involve multiple services. For example, every service may have 99.9% uptime, but if there are five services involved in each request, your uptime as a whole will only be 99.5%.[1] That might not seem like a huge difference, but it's actually an increase from 8 hours of downtime per year to 43!

Outright preventing failure across all your systems is simply not possible. There are too many components involved, the complexity is too high, and there is only so much you can invest into reliability without taking away time from user-facing features and other business needs. Since failure is inevitable, the best you can do is engineer your systems to handle failure gracefully. Handling failure gracefully means reducing the impact of failure as much as possible.

Consul can help reduce the impact of failure via its sidecar proxies. The three techniques looked at in this chapter are health checking, retries, and timeouts.

Health checking detects service failure and bypasses those services by routing to other healthy instances. There are two types of health checking, active and passive, that we'll examine. Retries are simply retrying requests when they fail in the hope that

---

1 $0.999^5 = 0.995$

it was a temporary issue. Finally, timeouts limit how long a service will wait for a response before assuming the request has failed.

First, we look at health checking.

Health Checking

Health checking is the process of determining the health of a service. A healthy service is responding to requests with minimal errors and in a reasonable amount of time. An unhealthy service is either fully down and not responding to any requests, or it is partially down and responding slowly or with an increased error rate.

If certain instances of a service are unhealthy, then Consul should route traffic away from the unhealthy instances to healthy instances (see Figure 8-1).

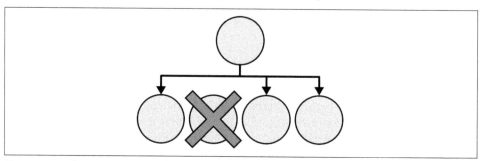

Figure 8-1. If instances are unhealthy, Consul shouldn't route requests to them

There are two ways to implement health checking: active and passive.

Active Versus Passive Health Checking

Active health checking is performing an action at a regular interval that tests if the service is healthy. For example, you may check if a port is open or a specific process is running, or you may call an endpoint. Many developers will implement an endpoint specifically to be called by active health checks.

If the test is successful, then the service is considered healthy. If the test is unsuccessful—for example, the port isn't open, the process isn't running, or the health check endpoint returns with an HTTP 500—the service is marked as unhealthy.[2]

Passive health checking relies on requests going to a service as part of its regular workload. Callers record all requests to a service, and if a certain number fail, the service is marked as unhealthy. For example, if you're making calls to the backend

---

2 Usually, the test will need to fail a certain number of times in a row before the service is considered unhealthy. This ensures that it isn't a temporary issue.

service and 9 out of 10 requests result in an HTTP 503 error, then you mark it as unhealthy.

Active health checks determine the status of the service for all callers. If the active check is failing, no services will route to it. In contrast, passive health checks only represent the status of the destination service for that specific caller. For example, if service A is calling endpoint A that's always failing, but service B is calling endpoint B that's always succeeding, then only service A will consider the service unhealthy.

 Passive health checking is also called *circuit breaking*. Like an electrical fuse that becomes overloaded and trips, the calling service will stop sending requests if too many fail.

Table 8-1 compares the two techniques.

Table 8-1. Comparison of active and passive health checking

	Active	Passive
Recency	Active checks are typically run on a regular schedule. This ensures the health status is always up to date, even when a service isn't receiving many requests as part of its normal workload.	The status of a service when using passive checks is based on previous requests. If a service hasn't received requests for some time, its health status might be outdated.
Clear status	Gives a single healthy or unhealthy status for the service.	If some endpoints are succeeding and others are failing, a passive health check won't give a clear status.
Representative of real traffic	Active checks usually test an endpoint specifically written for health checking. It's possible that this endpoint returns successfully while all the actual traffic to the service is failing.	Passive checks use the actual traffic to a service.
Locality aware	Active checks are usually run on the same node as the service. If there is a network partition between that node and other nodes, the active check will continue to pass, even though other services may not be able to reach this service.	Passive checks are run from the perspective of the calling service, so if there is a network partition, the calling service will correctly mark the upstream service as failing and stop calling it.

Active and passive health checks both have their advantages and disadvantages. That's why Consul uses a hybrid approach. Consul supports specifying active health checks that must be passing for a service to receive any traffic. But it also supports passive health checks by configuring the sidecar proxies to not route to upstream instances that are returning too many failures.

To further understand Consul's health checking, you'll configure active and passive checks for the backend service in the following sections.

Configuring Active Health Checks

Consul's active health checks are configured differently depending on whether you're running on Kubernetes or VMs.

On Kubernetes, Consul uses the data from Kubernetes readiness probes to determine the health status. Readiness probes are active health checks performed by the *kubelet*, a process that runs on every Kubernetes node. Since there's already a process performing active health checking, there's no need for Consul to do its own checking.

On VMs, there is no preexisting active health check system, so Consul performs the health checks itself.

First, I cover active checks on Kubernetes and then on VMs ("Configuring active health checks on VMs" on page 183).

Configuring active health checks on Kubernetes

Kubernetes calls its active health checks *probes*. Probes can be specified for each container in a pod. A Kubernetes probe can make HTTP requests—for example, to a /health endpoint—it can check if a port is open, or it can run a command inside the container.

There are three types of probes—startup, liveness, and readiness—but Consul only looks at readiness probes. In Kubernetes, readiness probes are used to indicate if a pod is ready to serve traffic. If a readiness probe fails for any container in the pod, Kubernetes will not route traffic to the pod. This is exactly what Consul cares about too: is a pod ready to receive traffic?

> Check out the Kubernetes documentation (*https://oreil.ly/rvZom*) to learn more about the other probe types.

Now that you understand how Consul uses Kubernetes probes, you're ready to test things out for yourself.

The backend service has a health endpoint at :7000/healthz built for health checking. This endpoint returns an HTTP 200 if the service is running. To test that Consul handles failure as expected, set a readiness probe to use port 9999. This will cause the probe to fail because there's nothing listening on that port.

Edit *backend-deployment.yaml* and add the readiness probe (see Example 8-1).

Example 8-1. backend-deployment.yaml

```
apiVersion: apps/v1
kind: Deployment
metadata:
  name: backend
  labels:
    app: backend
spec:
  replicas: 1
  selector:
    matchLabels:
      app: backend
  template:
    metadata:
      labels:
        app: backend
      annotations:
        consul.hashicorp.com/connect-inject: "true"
        consul.hashicorp.com/service-meta-version: "v1"
    spec:
      containers:
        - name: backend
          image: ghcr.io/consul-up/birdwatcher-backend:1.0.0
          env:
            - name: BIND_ADDR
              value: "0.0.0.0:7000"
            - name: TRACING_URL
              value: "http://jaeger-collector.default:9411"
          ports:
            - containerPort: 7000
          readinessProbe:
            httpGet:
              port: 9999 ❶
              path: /healthz ❷
            periodSeconds: 5
```

❶ Set the port to 9999 so that the probe will fail.

❷ The backend service has a health check endpoint at /healthz.

Apply the change:

```
$ kubectl apply -f backend-deployment.yaml

deployment.apps/backend configured
```

Wait a couple of seconds, and then check on the status of the backend pods:

```
$ kubectl get pods -l app=backend
```

```
NAME                         READY   STATUS    RESTARTS   AGE
backend-5846448594-hxrxc     2/2     Running   0          1h
backend-7895695fc5-r7bhj     1/2     Running   0          5s
```

You should see that there are two pods running, and one of the pods has its READY status set to 1/2. This means that only one of the containers in the pod is passing its readiness check (out of two).

 Kubernetes won't stop the old pod until the new pod is healthy.

Note the pod names in your cluster. In the preceding example, the ready pod is named backend-5846448594-hxrxc and the unready pod is named backend-7895695fc5-r7bhj.

Next, open up the Consul UI at *http://localhost:8500* (remember to be running `mini kube tunnel` or `kubectl port-forward`). You should see that Consul has marked the backend service as unhealthy (as shown in Figure 8-2).

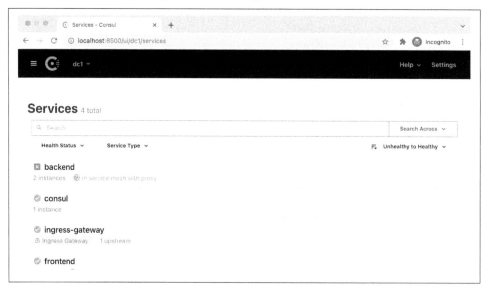

Figure 8-2. The backend service is marked as unhealthy in the Consul UI

If you click the backend service and then select the Instances tab (or navigate straight to *http://localhost:8500/ui/dc1/services/backend/instances*), you'll see that Consul has marked one of the instances as failing. In the example shown in Figure 8-3, it's named `backend-7895695fc5-r7bhj-backend`. This name should match the unhealthy pod you noted earlier.

Figure 8-3. The service instance corresponding to the unhealthy Kubernetes pod is marked as unhealthy in Consul's UI

Other Consul Checks

If you click on the service instance, you'll see all of its health checks:

Kubernetes Health Check
  The readiness probe status.

Destination Alias
  This check is registered on the sidecar proxy and will fail if any check on the service itself is failing. Under the hood, Consul treats the sidecar and the service as separate Consul services. Consul will use the status of the sidecar to determine service mesh routing, so this check ensures that the sidecar will fail if any of the service checks fail.

Serf Health Status
  The status of the local Consul client according to Serf gossip.

Proxy Public Listener
  Whether the sidecar proxy's port is open.

This shows that Consul has properly synced the readiness probe state from Kubernetes. Next, test out the service mesh routing and ensure traffic isn't being sent to the unhealthy pod. Load the Birdwatcher UI at *http://localhost:8080* and click Shuffle. You'll notice that the Host shown under Response is the name of the healthy pod as shown in Figure 8-4.

Figure 8-4. The Host data highlighted here shows the backend *pod that handled the* frontend *service's request. It should always be the healthy pod.*

This field is set to the name of the `backend` pod that serviced the request. Normally, Consul evenly distributes requests among pods. If Consul were ignoring the health status, you would expect it to route requests to both the unhealthy and healthy backend pods. However, if you continue to click Shuffle, you will see that Consul is always routing to the same pod. This proves that Consul is using the active health check status and not routing to the unhealthy pod.

Now, modify the `backend` deployment to use the correct port (7000) for the readiness probe:

```
# backend-deployment.yaml
apiVersion: apps/v1
kind: Deployment
metadata:
  name: backend
  # ...
spec:
```

```
# ...
template:
  # ...
  spec:
    containers:
      - name: backend
        # ...
        readinessProbe:
          httpGet:
            port: 7000
            path: /healthz
          periodSeconds: 5
```

Apply the change:

```
$ kubectl apply -f backend-deployment.yaml

deployment.apps/backend configured
```

Then wait for the rollout to be complete:

```
$ kubectl rollout status --watch deploy/backend

Waiting for deployment "backend" rollout to finish:
  1 old replicas are pending termination...
deployment "backend" successfully rolled out
```

If you open up the Consul UI, you'll see that the backend service is now marked as healthy. Kubernetes has stopped the old pod, and Birdwatcher will only be using the new backend pod.

Congratulations! You've successfully configured an active health check and tested that Consul is using it to route traffic. Skip ahead to "Passive Health Checks" on page 188 to learn about passive health checks.

Configuring active health checks on VMs

You can configure Consul to run active health checks against each service instance. The local Consul client on a node runs the checks and then reports the results to the Consul servers as shown in Figure 8-5.

> In the case of your Vagrant VM, since the services are registered with the Consul server, the Consul server will run the checks itself.

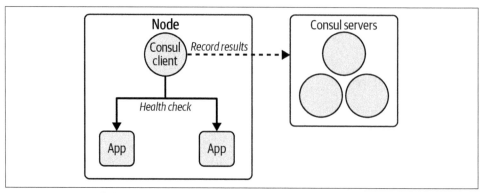

Figure 8-5. Consul clients health check services running on the same node and then send the results to the Consul servers

Consul supports many different ways to health check service instances. For example:

- Running a command and checking the result
- Making an HTTP request to the service instance and examining the response code
- Attempting to make a TCP connection

> If you want to learn more about the other types of checks, see Consul's health check documentation (*https://oreil.ly/U9hoK*).

To try out active health checking, configure a check for the backend service. Initially, create a check that you expect to fail. Then test whether Consul prevents routing to the backend service as expected. Next, reconfigure the check so it succeeds and test that traffic is routed normally.

You configure checks as part of the Consul service registration, so SSH into the VM so you can modify the service registration:

```
$ vagrant ssh
```

Edit */etc/consul.d/backend.hcl* and add a checks stanza as shown in Example 8-2.

Example 8-2. /etc/consul.d/backend.hcl

```
service {
  name = "backend"
  port = 7000
```

```
meta {
  version = "v1"
  prometheus_port = "20201"
}

connect {
  sidecar_service {
    port = 22000
    proxy {
      config {
        envoy_prometheus_bind_addr = "0.0.0.0:20201"
      }
    }
  }
}

checks = [
  {
    name = "Health endpoint" ❶
    http = "http://localhost:9999/healthz" ❷
    interval = "10s" ❸
    timeout = "1s" ❹
  }
]
}
```

❶ Each check must have a name.

❷ This is the URL Consul will use to perform the health check. Here, you're setting it to /healthz, the backend service's health check endpoint. You're also setting it to port 9999, which is the wrong port since the backend service is listening on port 7000. Using the wrong port will cause this check to fail.

❸ The interval is how often this check will run. A lower interval will mean the check is more up-to-date, but there will be more load on the service and Consul client.

❹ Timeout sets how long Consul will wait for a response before it considers the check a failure.

Reload Consul so it picks up the new health checks configuration:

```
$ consul reload
```

Now open up the Consul UI at *http://localhost:8500*. You will see the backend service listed as failing as shown in Figure 8-6.

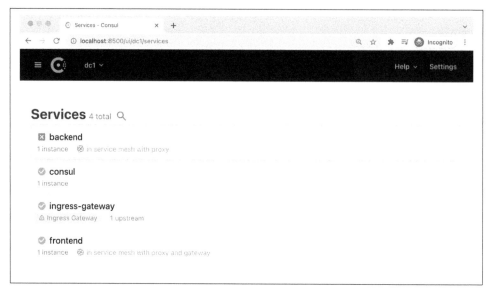

Figure 8-6. The backend service is shown as failing its health check

If you load the Birdwatcher UI at *http://localhost:8080*, you will now see an error because Consul has stopped routing to the unhealthy instance, and so there is no upstream for the frontend service to call (see Figure 8-7).

Figure 8-7. The Birdwatcher UI is getting an error because there are no healthy instances of the backend service to call

This test proves that active health checking is working as expected. If there were other healthy instances of backend running, Consul would route to those instead.

To get everything working again, you need to update backend's health check to use the correct port 7000.

Edit */etc/consul.d/backend.hcl* and change the URL of the check from `http://local host:9999/healthz` to `http://localhost:7000/healthz`:

```
service {
  name = "backend"
  ...
  checks = [
    {
      ...
      http = "http://localhost:7000/healthz"
      ...
    }
  ]
}
```

Reload Consul:

```
$ consul reload
```

Now look at the Consul UI again at *http://localhost:8500*. After a few seconds, Consul should list the backend service as healthy (see Figure 8-8).

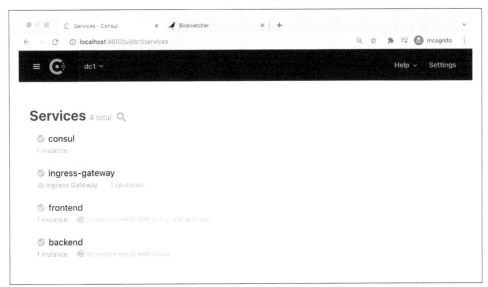

Figure 8-8. The backend service is now listed as healthy

You can see the healthy check by navigating to the health checks page for the back end service at *http://localhost:8500/ui/dc1/services/backend/instances/vagrant/backend/health-checks* or by clicking the backend service, selecting the Instances tab, and clicking the only instance in the list (see Figure 8-9).

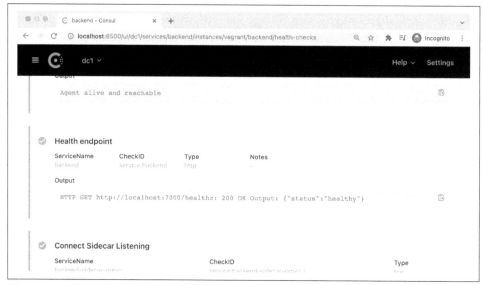

Figure 8-9. The output of the health check is shown in the Consul UI

If you refresh the Birdwatcher UI at *http://localhost:8080*, everything will be working normally again.

Congratulations! You've configured Consul's active health checks and tested that they work. In the next section, you'll learn how to configure passive health checks.

Passive Health Checks

Passive health checking in Consul is performed by the Envoy sidecar proxies.[3] Each sidecar proxy tracks the requests and responses to its upstream dependencies. As shown in Figure 8-10, if requests to an upstream service instance fail five times in a row (step 1), Envoy will mark the service instance as unhealthy and eject it from the routing pool (the list of service instances it can route to). It will then only route to the healthy instances (step 2). After 30 seconds, it will try to route to that service instance again. If the request succeeds, it will add it back to the routing pool. If it fails, it will wait even longer before trying it again.[4]

---

3 Envoy calls its passive health checking feature outlier detection (*https://oreil.ly/WAA4Q*).

4 By default, the maximum time it will wait between retrying is 300 seconds.

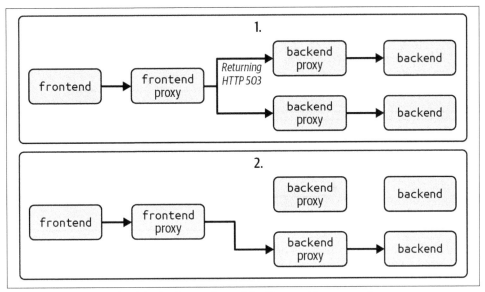

Figure 8-10. Passive health checks performed by Envoy sidecar proxies

 If the traffic is HTTP, Envoy considers any 5xx response code (that is, in the range 500–599) to be a failed request. With TCP traffic, Envoy considers a connection failure to be a failed request.

To better understand passive health checking, you can test it out yourself by causing a service instance to return HTTP 503s. After five failed requests, you should no longer see requests to that instance.

The Birdwatcher application supports specifying what percentage of requests should result in an error from the backend service. If you set this to 100% and make five requests (see step 1 in Figure 8-11), frontend's Envoy proxy should eject the backend service instance. Usually, Envoy would route around the unhealthy instance to a healthy instance. In this example, since you are only running one instance of the backend service, there will not be any healthy instances available, so Envoy will fail the request immediately (step 2).

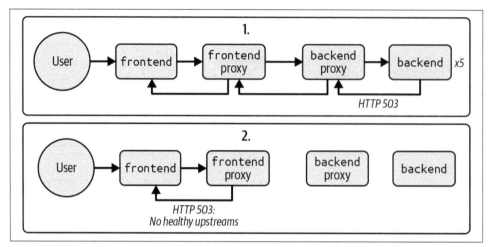

Figure 8-11. With only one instance of the backend service, if it fails passive health checks, Envoy fails the request immediately

To test this all out, open up the Birdwatcher UI at *http://localhost:8080* and slide the Error Rate setting to 100% as shown in Figure 8-12.

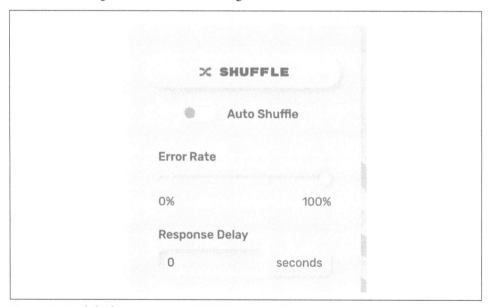

Figure 8-12. Slide the error rate to 100% to cause all requests to the backend service to return 503s

Now all requests to the backend service will result in HTTP 503. Click Shuffle six times. On the first five requests you will see:

```
Error: Received status code 503 from backend: "randomly generated error"
```

This error comes from the backend service returning an error due to the 100% error rate configuration.

But on the sixth and subsequent requests, you will see:

```
Error: Received status code 503 from backend: "no healthy upstream"
```

This error comes from frontend's proxy, which immediately returns a 503 because it has ejected the backend service instance from its routing pool. This proves that Envoy's passive health checking is working as expected. If you put the error rate slider back to zero and wait 30 seconds, Envoy will start routing to the service again.

This concludes the section on health checks. To summarize, there are two types of health checks: active and passive. Active health checking is performed by periodically testing to see if the service is healthy, and passive health checking is performed by the calling service by recording the success or failure of requests over time.

Through the exercises, you proved that Consul routes around instances that aren't passing their active or passive health checks, and when no instances are available, it fails the request immediately. In the next section, you'll learn about how retrying requests can further increase the reliability of your systems.

Retries

Even with health checks configured, there's still a chance your request may fail (see step 1 in Figure 8-13). This could be because of a random network issue or because a service has just become unhealthy. The easiest way to handle random failures like this is to simply retry your request. Consul supports configuring sidecar proxies to automatically retry requests (step 2). From the perspective of the calling services, they've only sent one request.

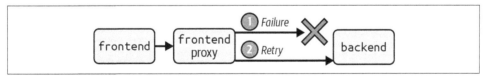

Figure 8-13. Consul retrying a request

To test Consul's retry capabilities, you can use the Birdwatcher app. In this exercise, you'll cause the backend service to error, and then configure frontend's proxy to retry requests.

If you've just completed the passive health checking exercise, wait until the Birdwatcher app is working (it should take 30 seconds for the backend service to be routable again).

Change the Error Rate slider of the Birdwatcher application to around 50%. Now when you click Shuffle, about 50% of the time you should see an error.

If you configure Consul to retry failed requests, you will be less likely to see an error. Given the percentage chance of an error and a maximum number of retries, you can calculate the chance of seeing an error. In this exercise, you'll set the maximum number of retries to five. This means with a 50% error rate, you should only reach the maximum number of retries (and therefore see an error) 3.1% of the time.[5]

You can configure retries using a new kind of config entry called a service router. Retries are configured from the perspective of the destination service, so you will create a service router for the backend service that's erroring.

If you're using Kubernetes, create a file named *backend-service-router.yaml* in *manifests/* with the contents from Example 8-3.

Example 8-3. backend-service-router.yaml

```
apiVersion: consul.hashicorp.com/v1alpha1
kind: ServiceRouter
metadata:
  name: backend
spec:
  routes:
    - destination:
        numRetries: 5 ❶
        retryOnStatusCodes: [503] ❷
```

❶ The maximum number of retries

❷ Which HTTP status codes will result in a retry

Apply the resource:

```
$ kubectl apply -f backend-service-router.yaml

servicerouter.consul.hashicorp.com/backend created
```

---

5 $0.5^5 = 0.03125$

And ensure it is synced:

```
$ kubectl get servicerouter backend

NAME       SYNCED   LAST SYNCED   AGE
backend    True     1s            1s
```

On VMs, SSH into the VM with `vagrant ssh` and then create a file named *backend-service-router.hcl* in the home directory ~ with the contents from Example 8-4.

Example 8-4. backend-service-router.hcl

```
Kind = "service-router"
Name = "backend"
Routes = [
  {
    Destination = {
      NumRetries = 5 ❶
      RetryOnStatusCodes = [503] ❷
    }
  }
]
```

❶ The maximum number of retries

❷ Which HTTP status codes will result in a retry

Write it to Consul:

```
$ consul config write backend-service-router.hcl

Config entry written: service-router/backend
```

Now that the service router config entry is in place on both Kubernetes and VMs, try clicking the Shuffle button again. You shouldn't see any more errors (unless you're very unlucky and hit that 3.1% chance).

You can see what's going on behind the scenes by looking at some of the traces in the Jaeger UI (*http://localhost:16686*) that have errors. For example, in the trace shown in Figure 8-14, you can see that the `frontend` proxy retried its request a second time after it failed initially.

Figure 8-14. The frontend proxy retried its request to the backend proxy after the first failure

Congratulations! You successfully considered Consul's retries and saw how without any changes to the underlying services, Consul can reduce the impact of random errors. Before you go about setting retries for all your services, however, please read the "Retries Considered Harmful?" sidebar. In the next section, you'll learn about another reliability technique: timeouts.

Retries Considered Harmful?

Retries can increase reliability in the face of temporary failures, but they can also cause upstream systems to be overloaded. For example, imagine there is an issue with the backend service that causes it to return HTTP 503s. If downstream services are configured to retry five times, the backend service will suddenly receive five times the traffic as every downstream service retries its requests.

This massive increase in load could actually further knock the backend service offline! Of course, with passive health checking in place, services would stop calling the backend service after five failures in a row, but passive health checking might not be triggered if the failures are intermittent.

Retries are a valuable tool, but they're not a silver bullet. I recommend *not* setting retries for your services except in situations when you know they're more helpful than harmful—for example, if you know your network is having temporary issues.

Timeouts

Another important technique for increasing reliability is timeouts. In a microservices architecture, there are many calls between services. Usually, the calling service has to wait for the response before it can respond back to *its* caller. In these systems, timeouts are vital.

Without timeouts, services will wait indefinitely for their upstream dependencies to respond. In most situations, upstream services respond promptly and so there are no issues. However, if an upstream service starts to run slowly, major problems can arise:

- Downstream services can run out of resources as more and more requests pile up, waiting for a response. This can take those services offline even though they weren't having any issues themselves.
- Load balancers can run out of connections. This can take unrelated services offline, even if they don't use the service that is running slow, because users can't get through the load balancers to access them.
- Users are stuck waiting for a response. In most cases, they'd prefer to see an error message quickly rather than wait indefinitely while the browser spins.

Timeouts solve these problems by failing the request after a defined period. You can implement timeouts in service code, but this has the same issues as implementing telemetry and security in service code. It makes more sense to leverage the service mesh.

To help test timeouts, the Birdwatcher application supports configuring a synthetic delay. Open up the Birdwatcher UI at *http://localhost:8080* and set Response Delay to 5 seconds. Also be sure to set Error Rate back to 0% as shown in Figure 8-15.

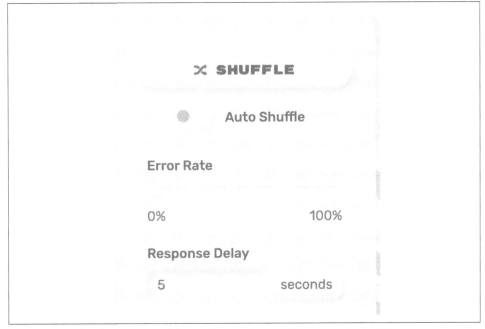

Figure 8-15. Set the response delay to 5 seconds and the error rate to 0%

This configuration will cause the `backend` service to wait five seconds before sending a response to the `frontend` service. Now click Shuffle. It should take five seconds before the next bird appears.

 The Birdwatcher UI shows how long the request took under the Duration header.

Imagine you want to set the timeout to one second. To do so you can use a service router config entry (the same one used for retries). Just like with retries, the timeout is set on the destination service, in this case, the `backend` service.

On Kubernetes, edit *backend-service-router.yaml* and add the `requestTimeout` configuration from Example 8-5 (you can also delete the retry settings).

Example 8-5. backend-service-router.yaml

```
apiVersion: consul.hashicorp.com/v1alpha1
kind: ServiceRouter
metadata:
  name: backend
spec:
  routes:
    - destination:
        requestTimeout: "1s"
```

Apply the new configuration:

```
$ kubectl apply -f backend-service-router.yaml
```

```
servicerouter.consul.hashicorp.com/backend configured
```

Ensure it is synced:

```
$ kubectl get servicerouter backend
```

```
NAME      SYNCED   LAST SYNCED   AGE
backend   True     1s            15m
```

On VMs, edit *backend-service-router.hcl* in the home directory (~) and add a `Request Timeout` value of `1s` to match Example 8-6 (you can also delete the retry settings).

Example 8-6. backend-service-router.hcl

```
Kind   = "service-router"
Name   = "backend"
Routes = [
```

```
{
  Destination = {
    RequestTimeout = "1s"
  }
}
]
```

Apply the config entry:

```
$ consul config write backend-service-router.hcl
```

```
Config entry written: service-router/backend
```

Now, whether on Kubernetes or VMs, go back to the Birdwatcher UI at *http://local host:8080* and click Shuffle again. The request should time out in one second with the error:

```
Received status code 504 from backend: "upstream request timeout"
```

If you look under the Response section, you'll see that the duration was close to one second, as shown in Figure 8-16.

Response

STATUS	DURATION
Error	998ms

HOST	VERSION
n/a	n/a

Figure 8-16. The response details show that the request had a duration of close to one second

Congratulations! You've successfully configured a timeout in Consul. In my experience, it is an absolute imperative to have timeouts for all your services. Otherwise a temporary latency issue can snowball into a full outage.

 One second might be too short a timeout in certain systems and too long a timeout in others. It all depends on your workload and the user experience you need to offer.

Remember to reset the response delay back to zero before moving on to the next chapter.

Summary

In this chapter, you learned about how to increase reliability using Consul. You learned about active and passive health checks and tested them with Consul. You saw how Consul reroutes traffic to avoid unhealthy instances and fails requests immediately when there are no healthy instances.

Then you learned how to configure retries using the service router config entry and tested it out on the Birdwatcher application. Finally, you saw how timeouts can reduce the impact of failure by ensuring services don't wait indefinitely for a response.

Building reliable systems is a difficult task. It requires good architecture and technology decisions, experience, and iteration. Consul is not a panacea for reliability; however, it can play an important part.

The next chapter covers traffic control: using Consul to control where traffic is routed.

## CHAPTER 9

Traffic Control

Traffic control is the ability to change where requests are routed. You already saw a glimpse of this when you tested health checking. When health checks failed, Consul configured the sidecar proxies to route to other instances. However, you didn't have control over this routing; it happened automatically.

This chapter will teach you how to use Consul to control traffic explicitly. The most common use case for traffic control is implementing deployment strategies like blue/green and canary. However, traffic control is also helpful for migration, when you're moving a service from one location to another, and for service refactoring, when you're splitting up or moving functionality between services.

The key benefit of Consul's traffic control features is that the underlying services don't need modification. As far as they're concerned, they're using the same URLs they've always been using, while in reality, the actual requests leaving the sidecar proxies might be different.

This chapter starts with a look at deployments.

Deployment Strategies

A successful deployment gets a new version of your service into the hands of real users without any downtime. The deployment process consists of two discrete steps:

1. Deploying the latest version
2. Routing traffic to the latest version

The two steps are sometimes referred to as *deploying* and *releasing*.

Figure 9-1 shows how deployment and routing can be separated. In step 1, only the v1 version of the service is running. In step 2, you deploy the v2 version, but it isn't receiving any traffic. Finally, in step 3, you swap traffic to v2.

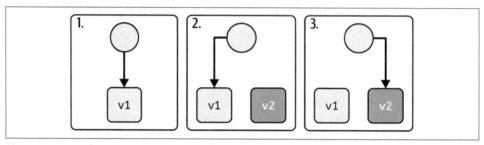

Figure 9-1. Deploying a new version of a service

Depending on your deployment strategy, the exact order of the steps can change. For example, some strategies will immediately route traffic to the v2 version, while others will route a small amount of traffic and slowly ramp up.

There are three primary deployment strategies:

- Rolling
- Blue/Green
- Canary

Rolling Deployments

In a rolling deployment, you slowly replace old instances with the newer version. At all times, traffic is routed equally across all running instances. For example, if two instances are running v1 and two are running v2, 50% of traffic will go to v1, and 50% of traffic will go to v2.

Figure 9-2 shows a rolling deployment. In step 1, there are three instances of v1. Each instance receives 33% of the traffic. In step 2, you start a new instance of v2. Traffic is routed equally, so 25% of traffic is now going to v2. In step 3, you replace an instance of v1 with a v2 instance. Now 50% of traffic is routing to v2. Finally, in step 4, you start a third instance of v2 and stop the remaining instances of v1. The deployment is now complete.

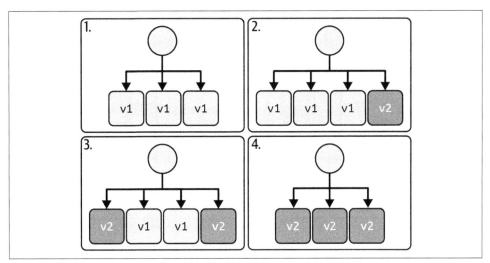

Figure 9-2. A rolling deployment

Rolling deployments are Consul's default strategy. Unless explicitly configured, Consul is not version-aware, so it routes to all service instances equally.[1]

Blue/Green Deployments

In a traditional blue/green deployment, there are two identical production environments. One environment is labeled green, and the other blue. There is only one environment that is actually servicing traffic at any one time. The other environment is either running on standby, or it's spun down. During a deployment, the new version is fully deployed to the inactive environment and then traffic is swapped over to that environment.

Figure 9-3 shows a blue/green deployment. In step 1, you've deployed v2 into the green environment, but all traffic is still being routed to the blue environment. In step 2, you swap traffic to the green environment, and 100% is now going to v2.

> Using two separate production environments for deployments isn't feasible in microservices architectures, because deployments happen so often across services that there would be too much routing churn. You can still use blue/green deployments with microservices, but you will deploy all versions to the same environment.

---

1 It's still up to your deployment orchestrator, like Kubernetes, to handle starting and stopping instances.

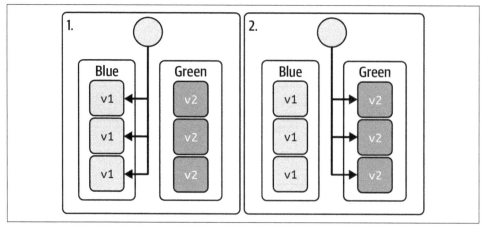

Figure 9-3. A blue/green deployment

Canary Deployments

A canary deployment is similar to a blue/green deployment in that you deploy the new version without it immediately starting to receive traffic. The main difference is that when it comes time to route traffic to the new version, traffic isn't swapped all at once. Instead, a small amount of traffic is routed at first to make sure it's working as expected.

> The "canary" deployment gets its name from the canaries coal miners in the early 1900s used as an early warning system for high carbon monoxide levels. The canary, in this case, is the small amount of traffic being sent to the newer version. If that traffic starts to result in errors, then it's an early warning that that version is bad.

Figure 9-4 shows the steps for a canary deployment. In step 1, only v1 is running. In step 2, you deploy one instance of v2, but only traffic with the HTTP header X-Version: v2 is routed to it. This allows you to test out the v2 version in production before it receives real traffic. In step 3, you slowly ramp up traffic until 10% of traffic is routed to v2. At this point, the canary pattern doesn't dictate how the rest of the rollout completes, but typically you'll use a rolling deployment to finish the rollout. In step 4, the rollout is complete.

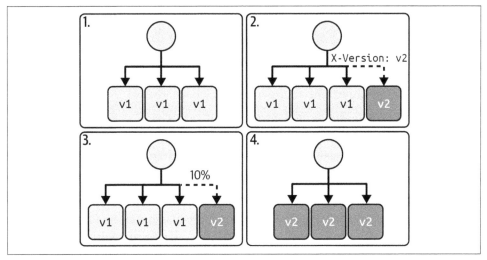

Figure 9-4. A canary deployment

When using the blue/green or canary deployment strategy, you need a way to control where traffic is routed. Traditionally this is the role of a load balancer, but a service mesh can perform this role more effectively.

Load Balancers Versus the Service Mesh

In many architectures, every service is fronted by a load balancer (see Figure 9-5), and deployment strategies are implemented by reconfiguring load balancers.

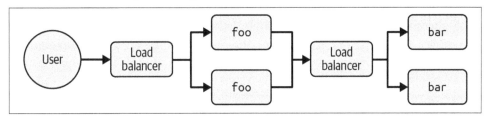

Figure 9-5. An architecture where every service has a load balancer

Load balancers have many disadvantages:

- They are usually difficult to configure dynamically, and you must configure each one individually.
- They add an extra network hop, which increases latency.
- They must be deployed independently from your services.

The Kubernetes Service resource improves upon load balancers because you don't have to deploy an actual load balancer, and it doesn't add an extra network hop, but it has its own limitations:

- It can't route traffic based on percentages. Each pod will receive the same amount of traffic. For example, with four pods, each pod will receive 25% of traffic.

- It can't route based on data in the request. For example, it doesn't support routing to certain pods based on an HTTP header.

- It doesn't support actual load balancing, only random routing. For example, if one pod is overloaded with requests, it will still receive the same amount of requests as other pods.[2]

With a service mesh, sidecar proxies act as miniature load balancers in front of every service instance. They are dynamically configurable, don't add additional hops,[3] are deployed alongside each service instance, and support all kinds of routing rules.

In the next section, you'll learn how to configure Consul's routing rules to implement your desired deployment strategy.

Traffic Control Config Entries

As mentioned previously, the default routing strategy for Consul is to route to all service instances equally. This is all that is needed for rolling deployments, but for blue/green and canary deployments, you need the finer-grained control provided by service resolver, splitter, and router config entries.

Service Resolvers

The service resolver config entry allows you to target different service versions by dividing the service instances into subsets. For example, as shown in Figure 9-6, the service resolver can identify and name two subsets of the backend service: v1 and v2.

To identify which instances belong in a specific subset, service resolvers allow you to filter through all instances and pull out the ones that match an expression. There are many attributes that can be filtered on (see the sidebar "Filter Language" on page 206), but the most common attributes are metadata key/value pairs.

---

2 Consul supports various load balancing algorithms using the service router config (*https://oreil.ly/ogDIh*).

3 Technically, they do add an additional hop, but it's within the same machine, so it adds microseconds rather than milliseconds (since the packets are just being moved by the Linux kernel, not across a network).

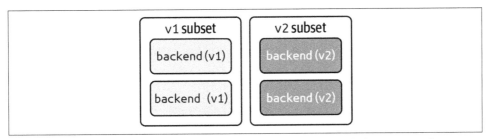

Figure 9-6. The backend service has four instances that are divided into two subsets

For example, when you registered the `backend` service in Consul, you added a metadata key: `version: v1`. In Kubernetes, this was done using:

```
annotations:
  "consul.hashicorp.com/service-meta-version": "v1"
```

And on VMs you used:

```
meta {
  version = "v1"
}
```

Imagine that you deploy a new version of the `backend` service with version set to v2. Using the expressions `Service.Meta.version == v1` and `Service.Meta.version == v2`, you can filter the instances into two subsets.

On Kubernetes, the service resolver config would be:

```
apiVersion: consul.hashicorp.com/v1alpha1
kind: ServiceResolver
metadata:
  name: backend
spec:
  subsets:
    v1:
      filter: 'Service.Meta.version == v1'
    v2:
      filter: 'Service.Meta.version == v2'
```

And on VMs, it would be:

```
Kind           = "service-resolver"
Name           = "backend"
Subsets = {
  v1 = {
    Filter = "Service.Meta.version == v1"
  }
  v2 = {
    Filter = "Service.Meta.version == v2"
  }
}
```

Once you've got your service subsets identified via a service resolver config entry, you're ready to configure how those subsets are routed to. By default, Consul will route to all service instances equally, regardless of subset, but you can customize this routing using service splitters and service routers.[4]

Service Splitters

A service splitter routes traffic to different subsets based on a percentage weight (see Figure 9-7).

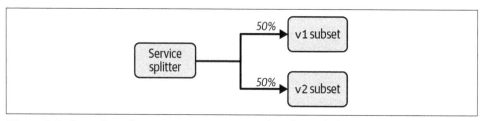

Figure 9-7. A service splitter routes percentages of traffic to different subsets

---

4 Note the name. Router, not resolver. The two config entries are easily mixed up!

On Kubernetes, a service splitter looks like:

```
apiVersion: consul.hashicorp.com/v1alpha1
kind: ServiceSplitter
metadata:
  name: backend
spec:
  splits:
    - serviceSubset: v1
      weight: 50
    - serviceSubset: v2
      weight: 50
```

On VMs, a service splitter looks like:

```
Kind = "service-splitter"
Name = "backend"
Splits = [
  {
    ServiceSubset = "v1"
    Weight        = 50
  },
  {
    ServiceSubset = "v2"
    Weight        = 50
  }
]
```

Service Routers

In Chapter 8, you used a service router to set retries and timeouts. You can also use a service router to match requests based on their attributes and route requests to certain subsets. Figure 9-8 shows a service router matching requests with the HTTP header X-Version set to v2 and routing them to the v2 subset. All other requests are routed to the v1 subset.

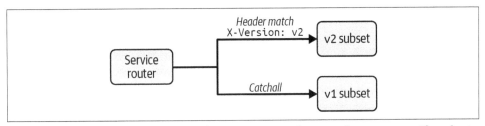

Figure 9-8. A service router can match on request attributes and route to specific subsets

For example, if you wish to route all requests that have the HTTP header X-Version: v2 to the v2 subset and all other requests to v1, you can use the config:

```
apiVersion: consul.hashicorp.com/v1alpha1
kind: ServiceRouter
```

```
metadata:
  name: backend
spec:
  routes:
    - match:
        http:
          header:
            - name: X-Version
              exact: v2 ❶
      destination:
        serviceSubset: v2 ❷
    - destination: ❸
        serviceSubset: v1
```

❶ exact means the header must match the string v2 exactly. Other supported settings include prefix, suffix, and regex.

❷ Route any matching requests to the v2 subset.

❸ A catchall: route all nonmatching requests to the v1 subset.

On VMs, that looks like:

```
Kind   = "service-router"
Name   = "backend"
Routes = [
  {
    Match = {
      HTTP = {
        Header = [
          {
            Name = "X-Version"
            Exact = "v2" ❶
          }
        ]
      }
    }
    Destination = {
      ServiceSubset = "v2" ❷
    }
  },
  {
    Destination = { ❸
      ServiceSubset = "v1"
    }
  }
]
```

❶ Exact means the header must match the string v2 exactly. Other supported settings include Prefix, Suffix, and Regex.

❷ Route any matching requests to the v2 subset.

❸ A catchall: route all nonmatching requests to the v1 subset.

Service routers are very flexible and can match requests based on HTTP or gRPC path, method, headers, or query strings.

Now you should be familiar with service resolvers, service splitters, and service routers. Service resolvers identify named subsets of service instances. These subsets are then referenced by service splitters to route certain percentages of traffic, and by service routers to route matching requests.

You'll learn how to use these config entries to implement canary deployments in the next section.

Canary Deployments with Consul

To experiment with the canary deployment strategy, you'll deploy a new v2 version of the Birdwatcher backend service. The backend v2 version is similar to the original v1 version, but it only returns canaries as shown in Figure 9-9.[5]

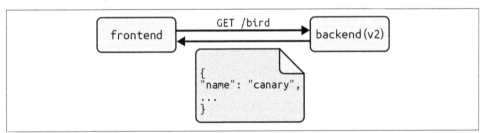

Figure 9-9. The v2 version of backend only returns canaries

The frontend service version doesn't change, but when backend v2 is rolled out, you'll only see canaries in the UI (as shown in Figure 9-10).

---

5 Too on the nose? Sorry. :)

Figure 9-10. When backend v2 is rolled out, Birdwatcher will only show canaries

The original version of the backend service will be referred to as v1 now.

The first step in a canary deployment is to ensure that traffic is only routed to the v1 version. When you eventually deploy the v2 version, you want to be able to control how much traffic it receives. By default, Consul routes all traffic equally between instances, so if you deploy backend v2 without any additional configuration, it will immediately receive traffic.

First, create a service resolver config entry to identify the v1 and v2 subsets. On Kubernetes, create a file *backend-service-resolver.yaml* in *manifests/* with the contents from Example 9-1.

Example 9-1. backend-service-resolver.yaml

```
apiVersion: consul.hashicorp.com/v1alpha1
kind: ServiceResolver
metadata:
  name: backend
spec:
```

```
    subsets:
      v1:
        filter: 'Service.Meta.version == v1'
      v2:
        filter: 'Service.Meta.version == v2'
```

Apply the resource:

```
$ kubectl apply -f backend-service-resolver.yaml
```

```
serviceresolver.consul.hashicorp.com/backend created
```

And ensure it is synced:

```
$ kubectl get serviceresolver backend
```

```
NAME       SYNCED   LAST SYNCED   AGE
backend    True     1s            1s
```

On VMs, vagrant ssh into the VM and create *backend-service-resolver.hcl* in ~ as shown in Example 9-2.

Example 9-2. backend-service-resolver.hcl

```
Kind         = "service-resolver"
Name         = "backend"
Subsets = {
  v1 = {
    Filter = "Service.Meta.version == v1"
  }
  v2 = {
    Filter = "Service.Meta.version == v2"
  }
}
```

Write it to Consul:

```
$ consul config write backend-service-resolver.hcl
```

```
Config entry written: service-resolver/backend
```

Now that the service resolver is in place, you can use the named subsets v1 and v2 in your service splitter and service router config entries. Create a service splitter that ensures 100% of traffic is routed to v1.

On Kubernetes, create the file *backend-service-splitter.yaml* as shown in Example 9-3.

Example 9-3. backend-service-splitter.yaml

```
apiVersion: consul.hashicorp.com/v1alpha1
kind: ServiceSplitter
metadata:
```

```
  name: backend
spec:
  splits:
    - weight: 100 ❶
      serviceSubset: v1
    - weight: 0
      serviceSubset: v2
```

❶   100% of traffic is routed to the v1 subset to start.

Apply the resource:

> `$ kubectl apply -f backend-service-splitter.yaml`

> `servicesplitter.consul.hashicorp.com/backend created`

And ensure it is synced:

> `$ kubectl get servicesplitter backend`

```
NAME      SYNCED   LAST SYNCED   AGE
backend   True     1s            1s
```

On VMs, create *backend-service-splitter.hcl* as shown in Example 9-4.

Example 9-4. backend-service-splitter.hcl

```
Kind = "service-splitter"
Name = "backend"
Splits = [
  {
    Weight        = 100 ❶
    ServiceSubset = "v1"
  },
  {
    Weight        = 0
    ServiceSubset = "v2"
  }
]
```

❶   100% of traffic is routed to the v1 subset to start.

And write the config entry:

> `$ consul config write backend-service-splitter.hcl`

> `Config entry written: service-splitter/backend`

Now Consul is configured to route all traffic to backend v1, and you can safely deploy backend v2 without it receiving any traffic. The current configuration is shown in Figure 9-11.

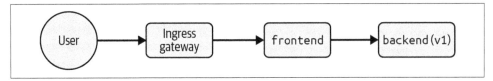

Figure 9-11. Currently, only backend v1 is running, and Consul is configured to route 100% of traffic to it

I'll cover how to deploy backend v2 first on Kubernetes and then on VMs ("Deploying backend v2 on VMs" on page 214).

Deploying backend v2 on Kubernetes

Usually, when deploying a new version on Kubernetes, you update the existing deployment with a new container image and let Kubernetes handle the rollout. Unfortunately, this causes Kubernetes to perform a rolling deployment, which isn't the strategy we want to use.

For a canary deployment, instead of updating the existing backend Deployment resource and letting Kubernetes perform the rollout, you will create a *new* Deployment resource for backend v2.

Create *backend-v2-deployment.yaml* as shown in Example 9-5.

Example 9-5. backend-v2-deployment.yaml

```
apiVersion: apps/v1
kind: Deployment
metadata:
  name: backend-v2
  labels:
    app: backend
spec:
  replicas: 1
  selector:
    matchLabels:
      app: backend
      version: v2
  template:
    metadata:
      labels:
        app: backend
        version: v2
      annotations:
        consul.hashicorp.com/connect-inject: "true"
        consul.hashicorp.com/service-meta-version: "v2" ❶
    spec:
      containers:
```

```
- name: backend
  image: ghcr.io/consul-up/birdwatcher-backend:1.0.0
  env:
    - name: BIND_ADDR
      value: "0.0.0.0:7000"
    - name: TRACING_URL
      value: "http://jaeger-collector.default:9411"
    - name: VERSION
      value: "v2"
  ports:
    - containerPort: 7000
  readinessProbe:
    httpGet:
      port: 7000
      path: /healthz
    periodSeconds: 5
```

❶ Note the metadata key version is set to v2.

Apply it:

```
$ kubectl apply -f backend-v2-deployment.yaml
```

And wait for the rollout to complete:

```
$ kubectl rollout status --watch deploy/backend-v2

deployment "backend-v2" successfully rolled out
```

Now backend v2 is running! Skip ahead to "Canary Deployment Continued" on page 217 to continue the exercise.

Deploying backend v2 on VMs

On VMs, you need to perform the following tasks to deploy the v2 version of the backend service:

1. Register the new backend v2 service with Consul.

2. Create new systemd unit files for backend v2 and its sidecar proxy.

vagrant ssh into the VM and create */etc/consul.d/backend-v2.hcl* (using sudo) with the contents from Example 9-6 to register the service instance with Consul.

Example 9-6. /etc/consul.d/backend-v2.hcl

```
service {
  name = "backend"
  id = "backend-v2"  ❶
  port = 7001
```

```
  meta {
    version = "v2" ❷
    prometheus_port = "20203"
  }

  connect {
    sidecar_service {
      port = 22001
      proxy {
        config {
          envoy_prometheus_bind_addr = "0.0.0.0:20203"
        }
      }
    }
  }

  checks = [
    {
      name = "Health endpoint"
      http = "http://localhost:7001/healthz"
      interval = "10s"
      timeout = "1s"
    }
  ]
}
```

❶ The ID of this service is backend-v2. Each service instance must have a unique ID, and there is already the original backend service instance running.

❷ Here, the meta key version is set to v2. This corresponds to the service resolver config.

The backend v2 service instance registration is almost identical to the backend v1 service instance registration. The only differences are separate ports and the version metadata key being set to v2 instead of v1.

Reload Consul so it picks up the new service instance:

```
$ consul reload

Configuration reload triggered
```

backend v2 is now registered with Consul. Next, create the systemd unit files to actually run the backend v2 service and its proxy.

Create /etc/systemd/system/backend-v2.service with the contents from Example 9-7 (you will need sudo).

Example 9-7. /etc/systemd/system/backend-v2.service

```
[Unit]
Description="Backend service v2"
Requires=network-online.target
After=network-online.target

[Service]
ExecStart=/usr/local/bin/backend
Restart=on-failure

Environment=BIND_ADDR=0.0.0.0:7001
Environment=TRACING_URL="http://localhost:9411"
Environment=VERSION=v2

[Install]
WantedBy=multi-user.target
```

Enable the service:

```
$ sudo systemctl enable backend-v2

Created symlink /etc/sys...
```

Start it:

```
$ sudo systemctl start backend-v2
```

Next, create its sidecar proxy's unit file at */etc/systemd/system/backend-v2-sidecar-proxy.service* (with sudo) as shown in Example 9-8.

Example 9-8. /etc/systemd/system/backend-v2-sidecar-proxy.service

```
[Unit]
Description="Backend v2 sidecar proxy service"
Requires=network-online.target
After=network-online.target

[Service]
ExecStart=/usr/bin/consul connect envoy -sidecar-for backend-v2 \
  -admin-bind 127.0.0.1:19003
Restart=on-failure

[Install]
WantedBy=multi-user.target
```

Enable the service:

```
$ sudo systemctl enable backend-v2-sidecar-proxy

Created symlink /etc/sys...
```

And start the proxy:

```
$ sudo systemctl start backend-v2-sidecar-proxy
```

backend v2 is now registered with Consul, and the service instance and its sidecar proxy are up and running.

Canary Deployment Continued

The current architecture is shown in Figure 9-12. There is one instance of backend v1 and one instance of backend v2, and Consul is routing 100% of traffic to backend v1.

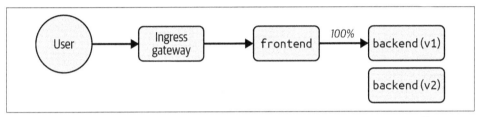

Figure 9-12. Both backend v1 and v2 are deployed, but only v1 is receiving traffic

If you navigate to *http://localhost:8080*, you should see regular birds and no canaries, as shown in Figure 9-13.

Figure 9-13. You should still be seeing regular birds because all traffic is routed to backend v1

Now you're ready to configure Consul to allow *some* traffic to backend v2. For this exercise, you will configure Consul to route traffic to v2 if it has the query parameter canary=true.[6] When frontend calls backend, it usually uses the path /bird. You will configure Consul so that if frontend uses /bird?canary=true, the request will be routed to backend v2.

To configure this rule, you'll use a service router config entry.

For Kubernetes, create a file *backend-service-router.yaml* to match Example 9-9 (if you already have this file from Chapter 8, modify it to match).

Example 9-9. backend-service-router.yaml

```
apiVersion: consul.hashicorp.com/v1alpha1
kind: ServiceRouter
metadata:
  name: backend
spec:
  routes:
    - match: ❶
        http:
          queryParam:
            - name: canary
              exact: "true"
      destination:
        serviceSubset: v2 ❷
```

❶   The match stanza specifies how to match the request.

❷   Route all matching requests to the v2 subset.

Apply the resource:

```
$ kubectl apply -f backend-service-router.yaml

servicerouter.consul.hashicorp.com/backend configured
```

And ensure it is synced:

```
$ kubectl get servicerouter backend

NAME      SYNCED   LAST SYNCED   AGE
backend   True     1s            1h
```

On VMs, create a file *backend-service-router.hcl* in ~ as shown in Example 9-10 (if you already have this file from Chapter 8, modify it to match).

---

6 Query parameters are key/value pairs passed at the end of the URL—for example, example.com? key=value&key2=value2.

Example 9-10. backend-service-router.hcl

```
Kind   = "service-router"
Name   = "backend"
Routes = [
  {
    Match = {  ❶
      HTTP = {
        QueryParam = [
          {
            Name = "canary"
            Exact = "true"
          }
        ]
      }
    }
    Destination = {
      ServiceSubset = "v2"  ❷
    }
  }
]
```

❶ The Match stanza specifies how to match the request.

❷ Route all matching requests to the v2 subset.

Write it to Consul:

```
$ consul config write backend-service-router.hcl

Config entry written: service-router/backend
```

Now that the router is configured, you can view it in the Consul UI's routing page for the backend service (*http://localhost:8500/ui/dc1/services/backend/routing*) as shown in Figure 9-14.

To test out the routing rule, navigate to *http://localhost:8080/?canary=true*. If everything worked, you should see a canary (as shown in Figure 9-10 on page 210)!

You may be wondering how setting the query parameter ?canary=true on requests to the *frontend* service can cause the routing rule for the *backend* service to be triggered. This is only possible in this example because the frontend service is programmed to read the canary query parameter and include it on its own call to the backend service.

In a real-life environment, you could use an HTTP header or a cookie that you've configured your services to forward.

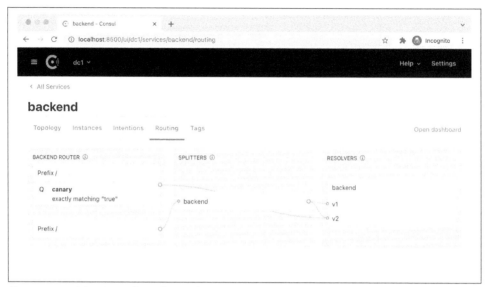

Figure 9-14. The Consul UI shows the routing rules

Click Shuffle a couple of times to ensure that everything is working correctly and you're only seeing canaries. In production, you'd likely look at your metrics and logs to confirm.

> If you remove the `?canary=true` query parameter you'll see that only regular birds are being returned for normal users.

Now that backend v2 looks good to you as a developer, it's time to route real user traffic to it. First, you'll route 25% of traffic.

On Kubernetes, edit *backend-service-splitter.yaml* and change the weights to have 25% of the traffic routed to v2:

```
apiVersion: consul.hashicorp.com/v1alpha1
kind: ServiceSplitter
metadata:
  name: backend
spec:
  splits:
    - weight: 75
      serviceSubset: v1
    - weight: 25
      serviceSubset: v2
```

Apply the new configuration:

```
$ kubectl apply -f backend-service-splitter.yaml

servicesplitter.consul.hashicorp.com/backend configured
```

On VMs, edit *backend-service-splitter.hcl* and change the weights to have 25% of the traffic routed to v2:

```
Kind = "service-splitter"
Name = "backend"
Splits = [
  {
    Weight       = 75
    ServiceSubset = "v1"
  },
  {
    Weight       = 25
    ServiceSubset = "v2"
  }
]
```

Apply the new configuration:

```
$ consul config write backend-service-splitter.hcl

Config entry written: service-splitter/backend
```

Navigate to the Birdwatcher app without the `canary` query parameter (*http://local host:8080*), and click Shuffle a couple of times. About one-quarter of the birds should be canaries.

Figure 9-15 shows the new system state.

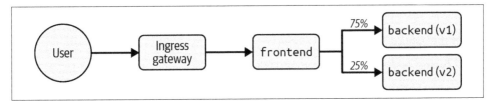

Figure 9-15. The new system state

In a real-life scenario you would now deploy more instances of backend v2, either in a rolling deploy or by spinning up more instances. You would then slowly route more traffic to v2. In this exercise, since there is only one instance of each service, you'll finish off the deployment by routing 100% of the traffic to v2.

On Kubernetes, edit *backend-service-splitter.yaml* and change the weights to have 100% of the traffic routed to v2:

```
apiVersion: consul.hashicorp.com/v1alpha1
kind: ServiceSplitter
metadata:
  name: backend
spec:
  splits:
    - weight: 0
      serviceSubset: v1
    - weight: 100
      serviceSubset: v2
```

Apply the new configuration:

```
$ kubectl apply -f backend-service-splitter.yaml
```

```
servicesplitter.consul.hashicorp.com/backend configured
```

On VMs, edit *backend-service-splitter.hcl* and change the weights to have 100% of the traffic routed to v2:

```
Kind = "service-splitter"
Name = "backend"
Splits = [
  {
    Weight        = 0
    ServiceSubset = "v1"
  },
  {
    Weight        = 100
    ServiceSubset = "v2"
  }
]
```

Apply the new configuration:

```
$ consul config write backend-service-splitter.hcl
```

```
Config entry written: service-splitter/backend
```

You should now only be seeing canaries in the Birdwatcher app, since 100% of the traffic is now being routed to backend v2. The canary deployment is now complete![7]

In this example, you saw how Consul's traffic control config can be used to implement a canary deployment. In the next section, you'll learn about other traffic control use cases.

---

7 Literally.

If you plan to implement regular canary deployments across all your services, you'll likely want to build some automation on top of Consul. Your automation could integrate with Consul's metrics and health checking to automatically roll back or finish the deployment, depending on the status of the new version. Start with Consul's HTTP API (*https://oreil.ly/OCHqi*) for config entries to see how you can manage config entries programmatically.

Other Traffic Control Use Cases

In addition to configuring deployment strategies, Consul's traffic control capabilities can also help with migration and service refactoring.

Migration is the process of moving a service from one location to another. One of the biggest hurdles to migrating a service is updating all of its callers to use the new address. With Consul, no updates are needed. Services can continue to use their existing URLs, and the sidecar proxies can automatically reroute traffic to the new addresses.

For example, imagine you are migrating the backend service from VMs to Kubernetes. Multi-cluster deployments will be covered in the next chapter, but for now, assume you have deployed the backend service into the Kubernetes cluster and you're ready to redirect traffic to it.

You can use a service resolver config entry to automatically redirect traffic:

```
Kind = "service-resolver"
Name = "backend"
Redirect {
  Service   = "backend"
  Datacenter = "kubernetes" ❶
}
```

❶ Consul calls each cluster a datacenter. This config will redirect the request to the Kubernetes datacenter.

Figure 9-16 shows how the service resolver will redirect traffic.

Service refactoring is another use case for traffic control. Imagine you're splitting up a large service into smaller microservices. Just like with a migration, you'd prefer not to update every caller. This is a perfect use case for the service router config entry. You can match on specific HTTP path prefixes and route those requests to a different service.

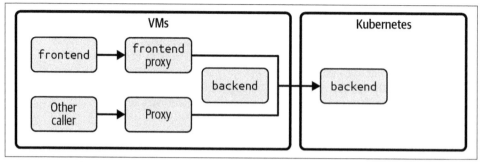

Figure 9-16. The service resolver will redirect all traffic to the backend service running in the Kubernetes cluster. The calling services can continue to use their original URLs.

For example, say you're moving the /bird endpoint on the backend service to a new service called birds. You can use the following service router config to match all /bird requests and route them to the new birds service:

```
apiVersion: consul.hashicorp.com/v1alpha1
kind: ServiceRouter
metadata:
  name: backend
spec:
  routes:
    - match:
        http:
          pathPrefix: /bird
      destination:
        service: birds
```

On VMs, it looks like:

```
Kind = "service-router"
Name = "web"
Routes = [
  {
    Match {
      HTTP {
        PathPrefix = "/bird"
      }
    }

    Destination {
      Service = "birds"
    }
  }
]
```

Summary

In this chapter, you learned how to control traffic using Consul. The first use case you tackled was deployment strategies.

You learned about the rolling deployment strategy, where you slowly spin down old instances as new instances come up. Then you learned about the blue/green strategy, where you bring up an equal number of new instances and then swap traffic over. Finally, you learned about canary deployments and how they give you even more control over exactly what traffic is routed to the new version.

Next, you were introduced to the config entries Consul uses for traffic control—service resolvers, splitters, and routers—and you used them to implement a canary deployment for the Birdwatcher application.

To wrap up, you learned about the traffic control techniques for managing migrations and service refactoring.

In the final chapter you'll learn about advanced Consul topics such as multi-cluster federation and ACLs.

## CHAPTER 10

Advanced Use Cases

This chapter examines advanced use cases for Consul and addresses areas I didn't have space to cover in depth.

Multi-cluster Federation

Consul supports being installed across multiple Kubernetes and VM clusters. Each cluster is called a Consul *datacenter*. Once installed across multiple clusters, you can easily and securely route traffic between clusters.

The Consul terminology for connecting multiple clusters is *federation*. Federation is accomplished through the use of Consul mesh gateways: essentially load balancers that are routable between clusters and through which all traffic flows, as shown in Figure 10-1.

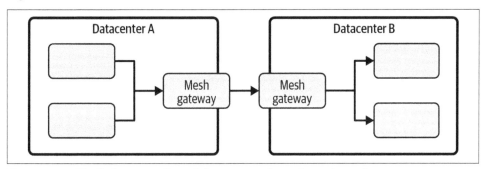

Figure 10-1. Traffic between Consul datacenters flows through mesh gateways

There are many use cases for multi-cluster federation:

- Connecting an on-premises datacenter running VMs with cloud-managed Kubernetes clusters
- Running multiple clusters for redundancy and configuring failover rules to route to other clusters in the event of failure
- Giving teams their own clusters (to reduce blast radiuses) while still enabling them to call services in other clusters

Intentions (Consul's authorization rules covered in Chapter 6) work across clusters, so you can ensure services are only talking to services in other clusters that they're authorized to.

To get started with multi-cluster federation, follow the instructions for Kubernetes (*https://oreil.ly/bcBHq*) or the WAN federation instructions (for VMs) (*https://oreil.ly/NQ2rU*) in the Consul documentation.

Consul API Gateway

Consul API gateways are a new add-on for Consul that work similarly to ingress gateways. They accept external traffic and route it securely into the mesh (as shown in Figure 10-2).

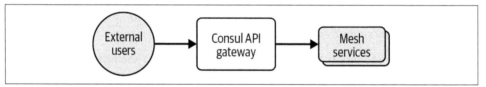

Figure 10-2. Consul API gateways accept external traffic and route it to mesh services

Consul API gateways are configured using a new collection of Kubernetes resources called the Gateway API (*https://oreil.ly/V39rk*). These resources are meant as a replacement for the Ingress resource.

> Currently (as of April 2022), Consul API gateways are only supported on Kubernetes, but support for Linux VMs is expected to be available before the end of the year.

Consul API gateways support using external certificates to terminate TLS traffic, unlike ingress gateways, and I anticipate more and more features will be added. I'm excited to watch the development of the API gateway because it allows you to use the same technology for all of your ingress traffic that you're already using for your service mesh and it promises to have more functionality than ingress gateways.

Visit the API gateway documentation (*https://oreil.ly/8cNBK*) for more information.

Terminating Gateways

Terminating gateways allow you to control access to services outside the service mesh. This is the opposite of ingress gateways that handle incoming traffic, as shown in Figure 10-3.

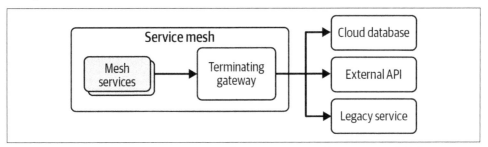

Figure 10-3. Terminating gateways control access to external services and APIs

By "services outside the mesh," I mean services that aren't running Consul sidecars. Many services can't run a sidecar, such as cloud databases (think Amazon Relational Database Service), external APIs, or even legacy services that you control but that are too difficult to add sidecars to.

To use a terminating gateway, register the external services with Consul. Mesh services then call those services like they're other mesh services. Under the hood, Consul handles routing those requests through the terminating gateway and then out to the external services.

At first, it might seem unnecessary to add a hop through a terminating gateway: why don't the mesh services route directly? The problem is that you need to open up firewall rules to all mesh services for the mesh services to route directly, and you can't control traffic with intentions. With a terminating gateway, you can use intentions to control which mesh services can talk to which external services, and then you only open up firewall rules for the terminating gateway.

Visit the terminating gateway documentation (*https://oreil.ly/5LnYq*) for more information.

HashiCorp Vault Integration

Vault (*https://oreil.ly/lVjdO*) is a secrets storage engine created by HashiCorp. Consul integrates with Vault by allowing you to configure Consul to use Vault's public key infrastructure (PKI) backend to generate TLS certificates. See the "Vault as a Connect CA" web page (*https://oreil.ly/hDY7K*) for more information.

Consul on Kubernetes can further integrate with Vault by allowing you to store the gossip encryption and control plane TLS certificates in Vault (the section "Securing Consul" on page 230 explains gossip encryption and control plane TLS). See the "Vault as the Secrets Backend" web page (*https://oreil.ly/VAj99*) for more information.

Connect Native

Connect Native is a method for integrating with the service mesh without running sidecar proxies. This is helpful for extremely high-performance applications that can't tolerate the overhead of a sidecar proxy. To use Connect Native, you will import or write your own library that uses the Consul API to discover the location of upstream dependencies and retrieve TLS certificates.

See the "Connect-Native App Integration" web page (*https://oreil.ly/VT8LT*) for more details.

Network Infrastructure Automation

Network Infrastructure Automation (NIA) is the ability to dynamically configure network infrastructure devices such as Cisco routers, F5 load balancers, and firewalls that aren't Consul aware. Consul-Terraform-Sync (CTS) is a Consul add-on that performs NIA by automatically configuring infrastructure devices in reaction to changes in Consul. For example, when a new service is registered, CTS can open up a firewall and add a routing config to a load balancer.

For more information on CTS, visit the Consul docs (*https://oreil.ly/trhD4*).

Securing Consul

The configuration of Consul I've shown in the book is not yet fully secure. Service-to-service traffic is encrypted, authorized, and authenticated, but Consul's internal communication is not yet secure. To fully secure Consul, you must enable ACLs, gossip encryption, and control plane TLS.

ACLs

ACLs are a method for controlling what actions an entity can perform (think authorization). All actions will require a token when you enable ACLs in Consul. Tokens have a list of actions that are allowed for that particular token.

For example, to register a service with Consul, you must have an ACL token with *write* permissions for that service's name. ACL permissions are specified using a policy language. The policy that gives write permissions for the service name frontend is specified as:

```
service "frontend" {
    policy = "write"
}
```

Enabling ACLs secures Consul because it prevents attackers from making unauthorized changes. For example, without ACLs, an attacker could register an address they control as the address for a service. Then Consul would route traffic to that address, allowing the attacker to capture internal traffic. Or an attacker could modify intentions and allow a service they control to make requests to any other service.

ACLs are turned on for Consul on Kubernetes via:

```
global:
  acls:
    manageSystemACLs: true
```

For VMs, review HashiCorp's "Secure Consul with Access Control Lists (ACLs)" tutorial (*https://oreil.ly/z4GuU*).

Gossip Encryption

Gossip encryption is used to encrypt Serf traffic between Consul servers and clients. Serf is Consul's decentralized membership and failure detection system and was discussed in Chapter 2. Serf traffic is encrypted using a symmetric encryption key shared by Consul servers and clients.

Gossip encryption helps secure Consul by ensuring that an attacker can't join a node to the cluster and discern the locations of other Consul nodes, or DDOS other nodes.[1] You can configure gossip encryption on Kubernetes via:

```
global:
  gossipEncryption:
    autoGenerate: true
```

On VMs, first generate a gossip encryption key using the `consul keygen` command, then set the `encrypt` key in every Consul server and client configuration file:

```
encrypt = "encryption-key-here"
```

Control Plane TLS

In Chapter 6, you learned how Consul encrypts service-to-service traffic using side-car proxies. However, by default, Consul doesn't encrypt *its own* traffic, as shown in Figure 10-4.

---

1 DDOS stands for "distributed denial of service" and is an attack method where huge amounts of network traffic is sent to other nodes to overwhelm them so they can't service regular traffic.

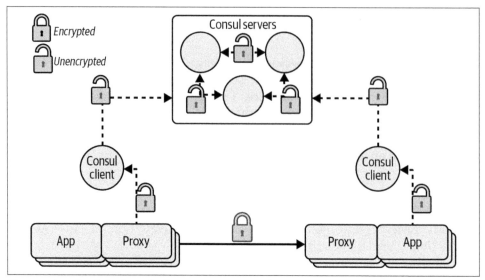

Figure 10-4. By default, only the communication between service mesh proxies is encrypted. Consul server-to-server, client-to-server, and proxy-to-client communication is not encrypted.

To encrypt Consul control plane traffic, TLS certificates must be distributed to each component.

Consul makes distributing certificates a bit easier via a mechanism called *auto-encrypt*, which will automatically distribute certificates to Consul clients (but not to servers or sidecar proxies).

On Kubernetes, you can configure control plane TLS with:

```
global:
  tls:
    enabled: true
    enableAutoEncrypt: true
```

On VMs, you must generate the certificates yourself and then distribute them to the Consul servers. The "Secure Consul Agent Communication with TLS Encryption" tutorial (*https://oreil.ly/CHD2h*) walks you through all the steps in detail.

Consul Enterprise

HashiCorp offers a paid version of Consul with enterprise-only features built for large companies. Consul Enterprise supports:

Admin partitions
> Big enterprises manage hundreds of Kubernetes and VM clusters. They typically have centralized platform teams responsible for this infrastructure. Admin partitions make it easier to manage each cluster by not requiring a set of Consul servers per cluster (like with multi-cluster federation). In addition, it makes it easier for the platform team to own the underlying Consul infrastructure.

Advanced Consul server features
> Automated backups, automated version upgrades, and a read-only mode to help with scaling.

To see the other features of Consul Enterprise, visit the Consul docs (*https://oreil.ly/U6uIF*).

HashiCorp Cloud Platform

HashiCorp Cloud Platform (HCP) offers HashiCorp tools as software as a service. HCP Consul is a hosted service on HCP that runs Consul servers for you and handles monitoring, backups, and upgrades. I'm a big believer in offloading as much operational overhead as possible, so if you're managing your own Consul installation, definitely take a look at HCP Consul.

See the HCP Consul web page (*https://oreil.ly/7LBCV*) for more details.

Amazon Elastic Container Service (ECS)

Consul can also run on Amazon ECS and provide the same service mesh functionality as on Kubernetes and VMs. To use Consul on ECS, you must include the sidecar proxy as a container in your ECS task, along with some other helper containers. You then connect with your upstream dependencies just like on VMs.

For more details, see the "AWS ECS" section (*https://oreil.ly/cZ1Ki*) of the Consul documentation.

Nomad

Nomad is a scheduler from HashiCorp that is similar to Kubernetes but is more lightweight and can run both containerized and noncontainerized workloads. Nomad integrates with Consul and handles automatically running sidecar proxies and registering services.

See the Nomad documentation (*https://oreil.ly/mBXPh*) for more details.

Conclusion

Congratulations, you've made it to the end! I hope you're excited about all the problems you can solve with Consul and that you now feel confident to deploy and operate Consul on your own. Thank you so much for reading!

 If you have more questions, please join the official *Consul: Up and Running* Discord server (*https://discord.gg/zxQcUVYKeS*) to get help from other readers and the author.

Common Errors

This is a list of errors you may encounter during the course of the exercises and how to fix them. Most of these errors are caused by the minikube cluster or Vagrant VM restarting.

Error	Solution

On Kubernetes you get an error:
`Internal error occurred: failed`
`calling webhook mutate-...`
`Post https://consul-controller-...`
`x509: certificate signed by`
`unknown authority`
	Restart the webhook certificate manager:
	`$ kubectl rollout restart \`
	`        deploy/consul-webhook-cert-manager -n consul`

On Kubernetes, services are not showing up in the UI.	Check the Consul client logs:
	`$ kubectl logs -l component=client -n consul`
	
	If you see `Error while renaming Node ID`, restart the Consul clients:
	`$ kubectl rollout restart ds/consul-client -n consul`

On Kubernetes, the state of the ingress gateway pod is stuck at Init.	Ensure minikube tunnel is running.

You get a `no healthy upstream` error when accessing the `frontend` service.	Restart the `frontend` service.
	On Kubernetes, run:
	`$ kubectl rollout restart deployment/frontend`
	
	On VMs, run:
	`$ sudo systemctl restart frontend`

Error	Solution

Metrics aren't showing up in Consul UI or Grafana.	Restart Prometheus.
	On Kubernetes, run:
	`$ kubectl rollout restart deploy/prometheus-server -n consul`
	
	On VMs, run:
	`$ sudo systemctl restart prometheus`

On VMs, you get an error accessing Birdwatcher:
`upstream connect error or disconnect/reset`
`before headers. reset reason: connection`
`failure, transport failure reason: TLS`
`error:..`
`CERTIFICATE_VERIFY_FAILED`
	Restart all proxies:
	`$ sudo systemctl restart ingress-gateway \`
	`    frontend-sidecar-proxy \`
	`    backend-sidecar-proxy \`
	`    backend-v2-sidecar-proxy`

If your error is not listed here, as a last resort you can uninstall Consul from your Kubernetes cluster or fully delete your minikube cluster or Vagrant VM and start over.

To uninstall Consul from your Kubernetes cluster, run:

> `$ consul-k8s uninstall`

To delete the minikube cluster, run:

> `$ minikube delete`

To delete the Vagrant VM, run:

> `$ vagrant destroy`

If you're still having issues, please join the official *Consul: Up and Running* Discord server (*https://discord.gg/zxQcUVYKeS*) to get help from other readers and the author.

Index

startup, 178
production deployment, multiple servers, 15
Prometheus
    configuring, 133-135
    data source, configuration, 144-145
    deploying, 133-135
    metrics
        Consul UI, viewing, 138-141
        ingress gateway, 135-138
prometheus_port metadata key, 136
PromQL (Prometheus Query Language), 145
provisioning VMs, 34-36
proxies
    Envoy, 2, 67
    HAProxy, 2
    NGINX, 2
    service instances, 3
    sidecar, 3, 14, 19
        configuring, 6
    traffic between services, 105
    transparent proxy, 58
proxy defaults, config entries, 82
public certificates, SPIFFE ID, 107
public ports, sidecar proxies, 71
public-key cryptography, 103

R
Raft, 16
readiness probes (Kubernetes), 178
reliability
    health checking
        active health checking, 176-188
        passive health checking, 176-177,
            188-191
    retries, 191-194
    service mesh and, 8-9
    timeouts, 194-198
replication, 17
request IDs, spans, 156
request retries, 191-194
resources in Kubernetes, custom, 83
retries, 191-194
rolling deployments, 200-201

S
security
    ACLs (access control lists), 230-231
    authentication, 100, 105-109
    authorization, 100, 109

castle and moat model, 98
certificate authority, 108
Control Plane TLS, 231-232
encryption, 100
firewalls, 98
gossip encryption, 231
identity and, 110
intentions, 109
IP addresses and, 110
man-in-the-middle attack, 101
Office of Personnel Management attack, 100
service auth, 101
service mesh and, 6
traffic between services and proxies, 105
traffic sniffing, 101
user auth, 101
zero trust network, 98-100
Serf, 18
service auth, 101
service failures, 18
service ID, SPIFFE, 107
service instances, 43
    filters and, 206
    sidecar proxies, 19
service intentions, config entries, 82
service mesh, 1
    certificates and, 6
    Connect Native, 230
    control plane, 2, 3
    example, 4-6
    features in combination, 10
    intentions, 69
    migration, no-downtime, 59
    observability, 7
    proxies, 2
        sidecar, 6
    reliability, 8-9
    security and, 6
    services, 47
        Kubernetes, 53-61
    traces, analysis, 171-172
    tracing
        enabling, 166-171
        on Kubernetes, 167-169
        on VMs, 169-171
    traffic and, 1
    traffic control, 9
    versus load balancers, 203-204
    when to use, 11

V

values.yaml file, 30, 76
VMs (virtual machines)
    active health checking, configuring, 183-188
    canary deployment, 214-217
    config entries, 84-85
    Consul operators, 41
    Consul servers, 16
    deploying Consul on, 34
        provisioning VM, 34-36
    deploying services on, 61-67
    Grafana installation, 143
    ingress gateways
        configuring, 91-93
        deploying, 79-82
    intentions
        application aware, 125
        configuring, 118-122
    Jaeger installation, 162-163
    service traces, 163-164
    services, registering, 64-66
    sidecar proxies, deploying, 67-71
    tracing service mesh, 169-171

W

workloads, network communication control, 1

Y

YAML, Kubernetes resources, 49

Z

zero trust network, 98-100

About the Author

Luke Kysow is a principal engineer at HashiCorp, where he works on Consul. He has extensive experience developing and operating applications in cloud and hybrid environments and has worked with many companies, large and small, to help them adopt Consul. He is also the cocreator of Atlantis, a popular open source Terraform CI/CD tool. He lives in Vancouver, Canada, with his fiancée, Isha.

Colophon

The animal on the cover of *Consul: Up and Running* is the crimson fruitcrow (*Haematoderus militaris*).

With a bushy crest reminiscent of that crowning a hoplite solider's helmet, this large, spectacularly colored bird is notable for the brilliant, glossy red plumage adorning its head, breast, and back. The plumage of females and immature males is paler than that of adult males, though equally striking.

Despite their name, these birds feed primarily on large insects in the subtropical and tropical lowland forests of Brazil, Guyana, and several neighboring countries of northern South America, where they spend the majority of their time perched high in the forest canopy. Fruit is, however, a supplement to their principal diet.

Relatively sluggish, with flight described as leisurely and undulating, or looping, the crimson fruitcrow is largely a solitary and quiet bird, though known to make an owl-like hooting or sharp barking call. Their population is believed to be in decline due to loss of habitat, but they have maintained a fairly extensive geographical range, a factor that has led to their current conservation status as being *of least concern*, or not threatened. Still, like all of the animals on O'Reilly covers, the crimson fruitcrow is vitally important to our world.

The cover illustration is by Karen Montgomery, based on an antique line engraving from Shaw's *General Zoology*. The cover fonts are Gilroy Semibold and Guardian Sans. The text font is Adobe Minion Pro; the heading font is Adobe Myriad Condensed; and the code font is Dalton Maag's Ubuntu Mono.

O'REILLY®

Learn from experts.
Become one yourself.

Books | Live online courses
Instant Answers | Virtual events
Videos | Interactive learning

Get started at oreilly.com.

Lightning Source UK Ltd.
Milton Keynes UK
UKHW031130170622
404579UK00002B/3